SpringerBriefs in Computer Science

Series Editors
Stan Zdonik
Peng Ning
Shashi Shekhar
Jonathan Katz
Xindong Wu
Lakhmi C. Jain
David Padua
Xuemin (Sherman) Shen
Borko Furht
V.S. Subrahmanian
Martial Hebert
Katsushi Ikeuchi
Bruno Siciliano

T0212818

For further volumes:
http://www.springer.com/series/10028

Mohamed M.E.A. Mahmoud
Xuemin (Sherman) Shen

Security for Multi-hop Wireless Networks

 Springer

Mohamed M.E.A. Mahmoud
Department of Electrical
  and Computer Engineering
Tennessee Technological University
Cookeville, TN, USA

Xuemin (Sherman) Shen
Department of Electronic
  and Computer Engineering
University of Waterloo
Waterloo, ON, Canada

ISSN 2191-5768   ISSN 2191-5776 (electronic)
ISBN 978-3-319-04602-0   ISBN 978-3-319-04603-7 (eBook)
DOI 10.1007/978-3-319-04603-7
Springer Cham Heidelberg New York Dordrecht London

Library of Congress Control Number: 2014930297

Printed on acid-free paper

Springer is part of Springer Science+Business Media (www.springer.com)

Preface

In multi-hop wireless networks, the mobile nodes should act as routers to relay others' packets. With multi-hop packet transmission, new applications can be enabled, the network performance can be enhanced, and the network can be deployed more readily in developing areas at low cost. However, the involvement of autonomous and self-interested nodes in packet routing can cause serious security vulnerabilities.

Selfish nodes will not relay others' packets because they consume the nodes' resources without direct benefits. They will also make use of the other nodes to relay their packets. This behavior will not only degrade the network connectivity and performance, but also introduce an unfairness problem. Moreover, some irrational attackers will launch *Denial-of-Service* attacks by involving their devices in routes with the intention of dropping packets. The presence of even a small number of attackers will result in repeatedly dropped packets. This can result in failure of the multi-hop communication or at least degrade the network performance in terms of throughput, delay, and packet delivery ratio. In addition, selecting good intermediate nodes to relay packets will have positive impact on route stability and packet delivery ratio. In this brief, we discuss efficient security protocols and schemes that can address these issues in multi-hop wireless networks.

In Chap. 1, we first discuss in detail the security issues addressed by this brief, and then we discuss the challenges for securing the multi-hop wireless networks. Finally, we give an overview of the proposed protocols and schemes to secure the multi-hop wireless networks. In Chap. 2, we present the considered system model including the network and communication models and the threat and trust models. In Chap. 3, we present an efficient incentive scheme to stimulate the selfish nodes to relay others' packets. The scheme uses payment system to charge the nodes that send packets and reward those relaying them. In addition to stimulating cooperation, the scheme can enforce fairness by rewarding the nodes for the consumed resources in relaying others' packets. It can also regulate packet transmission because the nodes pay to get their packets delivered. We first develop a payment model that is specifically tailored to our system model. Then, we discuss an efficient communication protocol to secure the payment with limited use of public key cryptography. We also discuss a mechanism for submitting the payment data to an off-line trusted party and clearing the payment with minimum overhead. Finally, analysis and measurements will be given to verify our proposals.

In Chap. 4, we first discuss a trust system to evaluate the nodes' competence and reliability in relaying packets in terms of multi-dimensional trust values. Then, we present a mechanism to identify the irrational attackers that drop packets intentionally. A node is identified as malicious once its packet dropping rate measured by the trusted system exceeds a threshold. We also discuss two routing protocols for establishing stable and reliable routes in multi-hop wireless networks. The protocols ensure relaying the packets by those highly trusted nodes having sufficient energy to minimize packet dropping probability. We will show that the integration of routing protocols with the trust and payment systems not only stimulates the nodes to relay others' packets but also maintains route stability and reports correct battery energy level. This is because any loss of trust will result in loss of future earnings. In addition, analysis and simulation results will be given to evaluate these protocols and mechanisms. Finally, we draw some conclusions and future research directions in Chap. 5.

Cookeville, TN, USA   Mohamed M.E.A. Mahmoud
Waterloo, ON, Canada   Xuemin (Sherman) Shen

Contents

List of Figures

List of Tables

Chapter 1
Introduction

In multi-hop wireless networks (MWNs), when a mobile node needs to communicate with a remote destination, it relies on the other nodes to relay its packets [1–4]. The network nodes commit bandwidth, data storage, CPU cycles, battery energy, etc, forming a pool of resources that can be shared by all of them. The utility that the nodes can obtain from the pooled resources is much larger than that they can obtain on their own. MWNs can enable new applications and enhance the network performance [5–7]. The multi-hop packet relay can extend the communication range with using limited transmit power because packets are transmitted over shorter distances. It can also improve the area spectral efficiency and the network throughput and capacity [8]. Moreover, MWNs can be deployed more readily and at low cost in developing and rural areas. We consider the civilian applications of multihop wireless networks, where the network has long lifetime and the mobile nodes have long-term relations with the network. MWNs can implement many useful applications such as data sharing and multimedia data transmission [9], [10]. For example, users in a residential neighborhood or a university campus may carry different wireless-enabled devices such as PDAs, laptops, tablets, cell phones, etc. They can establish a network to communicate, distribute files, and share information. However, the involvement of autonomous and self-interested nodes in packet relaying, makes MWNs vulnerable to serious security vulnerabilities that may hinder their practical implementation.

1.1 Security Issues

In military and disaster-recovery applications, the nodes' behavior can be highly predictable because the network is closed and the nodes belong to and are controlled by a single authority. The nodes also pursue a common goal. However, the nodes' behavior is unpredictable in civilian applications because they are typically autonomous and self-interested and may belong to different authorities.

M.M.E.A. Mahmoud and X. Shen, *Security for Multi-hop Wireless Networks*,
SpringerBriefs in Computer Science, DOI 10.1007/978-3-319-04603-7_1,
© The Author(s) 2014

Most existing protocols, such as DSR [11], assume that the nodes are willing to relay others' packets. This assumption can be acceptable for disaster recovery and military applications but it cannot be guaranteed in civilian applications. This is because the nodes aim to maximize their welfare and minimize their contributions. Although the proper network operation requires the nodes to cooperate in relaying others' packets, this consumes their valuable resources such as battery energy and computing power without any direct benefits. Therefore, in civilian applications, selfish nodes will not be voluntarily interested in cooperation without sufficient incentive and make use of the cooperative nodes to relay their packets. This behavior will not only degrades the network performance but also creates fairness problem. The fairness issue arises when the selfish nodes take advantage of the cooperative nodes without any contributions to them, and the cooperative nodes are unfairly overloaded because the network traffic is concentrated through them. When more nodes are cooperative in relaying packets, the routes are shorter, the network connectivity is more, and the possibility of network partition is lower. The selfish behavior also degrades the network performance significantly, which may result in failure of the multi-hop communication. The evaluations given in references [12, 13] have proven that if 10 to 40 % of the nodes behave selfishly, the average throughput degrades by 16 to 32 %, and the delay increases linearly with the percentage of selfish nodes.

Moreover, packets may be dropped and a route is broken when an intermediate node moves out of the radio range of its neighbors in the route. However, some malicious nodes (or irrational packet droppers) launch *Denial-of-Service (DoS)* attacks by actively involving themselves in communication routes with the intention of dropping the packets to disrupt the communication. These malicious nodes can be compromised or malfunctioned nodes having faulty hardware or software. In addition, since the nodes are equipped with different hardware capability, the nodes having larger resources, such as CPU speed, buffer size and battery energy, can perform packet relay more successfully than others. For example, PDAs may not be able to relay packets effectively due to the scarcity of resources. Some nodes may break routes because they do not have sufficient energy to relay the source nodes' packets to keep the routes connected. Because of this uncertainty in the nodes' behavior, randomly selecting the intermediate nodes may degrade the routes' stability. It will also endanger the reliability of data transmission and considerably degrade the network connectivity and performance in terms of packet delivery ratio and throughput [12]. It may also cause the multi-hop communication to fail. Only one intermediate node can break a route, and the presence of even a small number of attackers or incompetent nodes can repeatedly drop packets and break routes. When a route is broken, the nodes have to rely on cycles of time-out and route discoveries to re-establish the route. These route discoveries may incur network-wide flooding of routing requests that consume a significant amount of the network's resources. They will also increase the packet delivery latency. Hence, in order to maintain continuous traffic flow, stable and reliable routes should be established by selecting good intermediate nodes.

1.2 Motivations and Challenges

In the literature, reputation-based and incentive schemes [14–18] have been proposed to enforce and stimulate node cooperation, respectively. For reputation-based schemes, packet relay is an obligation. Each node should monitor the transmissions of its neighbors to make sure that the neighbors relay others' packets and thus selfish nodes can be identified and punished. Since packets can be dropped due to non-malicious reasons such as mobility, collision, and impaired channel, each node should measure the rates of dropping packets by the other nodes in terms of reputation values. When a node $\mathcal{N}_A$ transmits a packet to the next node in the route $\mathcal{N}_B$ to relay to $\mathcal{N}_C$, $\mathcal{N}_A$ has to overhear the medium to make sure that $\mathcal{N}_B$ relays the packet. If $\mathcal{N}_A$ does not overhear the packet transmission, it assumes that $\mathcal{N}_B$ has dropped the packet. $\mathcal{N}_A$ increases the reputation value of $\mathcal{N}_B$ when it observes a packet transmission, and decreases the reputation value when it observes a packet drop. In other words, a node's reputation value is improved when it relays packets and degraded when it drops packets. Once a node's reputation value degrades to a threshold, the node is identified as selfish and punished.

For incentive schemes, relaying other nodes' packets is a service not an obligation. The schemes use credits (or micropayment) to motivate the selfish nodes to collaborate by making cooperation more beneficial than behaving selfishly. The nodes earn credits for relaying others' packets and spend them to get their packets delivered. In addition to stimulating cooperation, incentive schemes can enforce fairness, discourage *Message-Flooding* attacks, and regulate packet transmission. Fairness can be enforced by rewarding the nodes that relay packets and charging those that send packets. For example, the nodes situated at the network center relay more packets than the other nodes because they are more frequently selected by the routing protocol. Since the source nodes pay for relaying their packets, the incentive schemes can also regulate packet transmission and discourage *Message-Flooding* attack. In this attack, the attackers send bogus messages to deplete the intermediate nodes' resources. Incentive schemes can also be used for charging the future services of the mobile networks because the mobile nodes may roam among different foreign networks, and thus a trusted central unit may not be involved in the communication sessions [19], i.e., a mobile node can pay different network operators without contacting a distant home location register.

However, the reputation-based schemes suffer from the following essential issues that discourage using them for the civilian applications of MWNs.

1. Monitoring by overhearing the medium may not work in a number of situations: (1) When $\mathcal{N}_B$ relays a packet to $\mathcal{N}_C$, $\mathcal{N}_A$ cannot overhear the transmission because of the packet collision due to another concurrent transmission in its neighborhood [20]; and (2) Since $\mathcal{N}_A$ can know if $\mathcal{N}_B$ has relayed a packet but cannot know if $\mathcal{N}_C$ received it, $\mathcal{N}_B$ can save its power and circumvent the monitoring technique if $\mathcal{N}_A$ is closer than $\mathcal{N}_C$ by adjusting its transmission power such that the signal is strong enough to be overheard by the monitoring node $\mathcal{N}_A$ but too weak to be received by the true recipient $\mathcal{N}_C$ [20].

2. The channel overhearing monitoring technique is not energy-efficient for transmitters because the full power transmission is used instead of adapting the transmission power according to the distance separating the transmitter and the receiver to enhance the scheme's effectiveness by enabling more neighboring nodes to overhear the packet transmission [21]. Furthermore, the directional antennas [22] that can improve the network capacity due to reducing the interference area make monitoring difficult.

3. Reputation-based schemes cannot enforce fairness because they can not compensate the highly contributing nodes. They force the nodes to serve the network without any benefits and punish them when they fail to cooperate no matter how they have previously contributed to the network. For example, although the nodes situated at the network center relay more packets than the other nodes because they are selected more by the routing protocols, they are not compensated.

4. These schemes may not be accurate in detecting selfish nodes and may falsely accuse honest nodes because it is difficult to differentiate between a node's unwillingness and incapability to cooperate, e.g., due to low resources, bad channel, and network congestion. In order to precisely identify the selfish nodes, the nodes' behavior should be observed over a long time and different sessions but the schemes may not have sufficient time to judge a node's behavior precisely due to the node mobility. Moreover, the assumption that a node's transmission can be received by all the nodes in its neighborhood cannot be ensured, e.g., due to packet collision [20].

Although the incentive schemes are attractive for the civilian applications of the MWNs, they suffer from the following important issues.

1. Micropayment schemes have been originally designed for Internet electronic commerce applications to take advantage of the high volume of viewers by offering content for low price. Examples of the applications include buying data or news, listening to a song, playing an online game, and reading an article in a journal [23]. In order to efficiently implement micropayment in multi-hop wireless networks, the payment model should be specifically tailored to cooperation stimulation in MWNs and should consider the differences between Web-based applications and cooperation stimulation.

2. The non-reputation property of the public key cryptography can help secure the payment effectively. However, the extensive use of this cryptography is too resource consuming for mobile devices because it requires too many computations. Moreover, secure public-key cryptosystems usually have long signature tags that increase the packet overhead. Securing the payment without using public key cryptography or even with limited use of this cryptography is a challenge.

3. Since a trusted party is not involved in the communication sessions, the nodes usually compose undeniable proofs of relaying others' packets, called receipts, and submit them to an off-line trusted party (Tp) to claim the payment. The receipts' size is large because they carry security proofs, e.g., signatures, to secure the payment, which significantly consumes the nodes' resources and the

available bandwidth in submitting them. Tp has to apply a large number of cryptographic operations to verify the receipts. Reducing the receipts' submission and processing overhead is essential for the effective implementation of the incentive scheme. Submitting a large number of receipts imposes significant communication and processing overhead and implementation complexity. Moreover, since a transaction (relaying packets) value may be very low, the scheme uses micropayment, and thus a transaction's overhead in terms of submitting and clearing the receipts should be much less than its value.

4. The incentive schemes assume that the nodes are rational in the sense that they will relay the packets faithfully when this is more beneficial than dropping the packets. This assumption cannot be guaranteed in MWNs because it ignores the irrational attackers and malfunctioned nodes. It also ignores the fact that the nodes loaded with low hardware resources will drop packets because they lack the CPU cycles or the buffer space to relay the packets. In IP networks, the malfunction of the network equipments is an important source for the network unavailability [24]. In other words, *using credits alone is not enough to ensure that the nodes will not drop packets and the routes are stable.*

1.3 Overview on the Proposed Protocols/Schemes

In this brief, we first develop a payment model that takes into account the features of cooperation stimulation in order to improve the effectiveness of using micropayment in MWNs. Second, based on this payment model, we propose a fair, efficient and secure incentive scheme. For fair charging policy, both the source and the destination nodes are charged when the two nodes benefit from the communication. To reduce the overhead, we discuss a communication protocol that can secure the payment with limited use of public-key cryptography. Public-key cryptography operations are used only for the first packet in a series and the efficient hashing operations are used for the next packets, so that the overhead of the packet series converges to that of the hashing operations. Hash chains and keyed hash values are used to ensure payment non-repudiation and thwart free riding attacks. The protocol is called ESIP (Efficient and Secure Incentive Protocol).

Then, we will propose a scheme to reduce the overhead of submitting/processing the payment receipts. This scheme is called RACE (**R**eport-based p**A**yment s**C**hem**E**). Instead of submitting large-size payment receipts, the nodes submit brief payment reports to Tp and temporarily store undeniable cryptographic tags called *Evidences*. The reports contain the alleged charges and rewards without any cryptographic proofs. RACE can verify the reports' credibility by investigating the consistency of the reports, and clear the payment of the fair reports with almost no cryptographic operations or computational overhead. For the cheating reports, RACE requests the *Evidences* to identify and evict the cheating nodes that submit incorrect reports. If cheating actions are not too often, RACE can significantly reduce the required bandwidth and energy for submitting the payment data and clearing the payment. Instead of requesting the *Evidences* from all the nodes participating in the cheating reports, RACE can identify the cheating nodes with requesting few *Evidences*. Moreover, *Evidence* aggregation technique is used to reduce the *Evidences'* storage area. In RACE, *Evidences* are submitted and Tp applies cryptographic operations to verify them only in case of cheating, but the nodes systematically submit security tokens, e.g., signatures, and Tp always applies cryptographic operations to verify them in the existing receipt-based schemes.

We also develop a trust/reputation system to continuously evaluate the nodes' packet-relay success rates in terms of trust/reputation values. Trust systems have been used in a wide range of applications, such as public key authentication, electronic commerce, supporting decision making, etc [25–27]. In MWNs, using trust system is essential to assess the nodes' trustworthiness, competence, and reliability in relaying packets. A node's trust value is defined as the degree of belief in the node's behavior, i.e., the probability that the node will behave as expected. The trust values are calculated from the nodes' past behaviors and used to predict their future behavior. For example, there is a strong belief that a node will break a route if it has broken a large percentage of routes in the past. Most of the existing trust systems compute a single trust value for each node. However, a single measure may not be expressive enough to adequately depict a node's trustworthiness and competence. We propose a trust system that maintains multi-dimensional trust values for each node to evaluate the node's behavior from different perspectives. Multi-dimensional trust values can better predict the nodes' future behavior, and thus help make smarter routing decisions. In our trust system, the nodes that frequently drop packets, break routes, or are not active in relaying packets have low trust values. Moreover, for the efficient implementation of the trust system, Tp computes the trust values by processing the payment reports instead of using the traditional medium overhearing technique. The reports can be processed to extract financial information to clear the payment, as well as contextual information to update the trust values. Our approach is based on this observation: *if a message is received by node $\mathcal{N}_i$, this means that all the nodes from the source to $\mathcal{N}_{i-1}$ have successfully relayed the message.*

In addition, we propose a mechanism that uses the nodes' trust/reputation values to identify the malicious node that frequently drop packets and break routes. It may be impossible to know whether a packet is dropped for malicious or nonmalicious reasons, e.g., due to faulty nodes and bad channel condition; therefore, it is necessary to measure the nodes' frequency of dropping packets to figure out if dropping packets is genuine or temporary action. If packet dropping is a genuine action, the malicious node's reputation value degrades to a threshold. In this case, the node is identified as malicious and excluded from the network because it becomes a threat to the network proper operation. This mechanism is called TRIPO (A mechanism for Thwarting Rational and Irrational Packet drOpping attacks). Similar to the existing incentive schemes, the nodes are stimulated and not forced to relay others' packets, and similar to the existing reputation-based schemes, consistently dropping packets and breaking communication routes is an obvious abuse due to disrupting the routing process. *By this way, credits are used to stimulate the selfish nodes (or rational attackers) to relay packets and behave faithfully, and reputation values are used to force the malicious nodes (or irrational attackers) to behave rationally to avoid eviction.*

Finally, thwarting selfish and malicious nodes is not sufficient to ensure route stability which requires selecting highly trusted nodes having sufficient battery energy. The nodes will have different packet-relay success rates expressed by their trust values. We discuss two routing protocols to establish stable and reliable routes by selecting good intermediate nodes. The routing protocols establish stable and reliable routes through the highly trusted nodes having sufficient residual battery energy in order to minimize the packet dropping probability. Establishing good routes can significantly improve the network performance in terms of packet delay, throughput, and packet delivery ratio. *By this way, the integration of payment/trust systems with the routing protocol not only stimulates the nodes to relay others' packets but also to maintain route stability and report correct battery energy level.* This is because any loss of trust will result in loss of future earnings.

Extensive security analysis demonstrate that our proposals can secure the payment and trust calculation, and precisely identify the malicious and the cheating nodes with almost no false accusations. Moreover, our measurements demonstrate that the incentive scheme requires much less overhead than the signature-based protocols because the lightweight hashing operations dominate the nodes' operations. For a series of 13 packets, the simulation results demonstrate that ESIP requires less than 10% of the computation time and energy required in the DSA-based incentive schemes. Moreover, the average packet overhead is less than those of the public-key-based protocols with very high probability.

Our simulation results demonstrate that RACE requires much less communication and processing overhead than the receipt-based schemes with acceptable payment clearance delay and storage area. Moreover, our evaluations demonstrate that RACE can secure the payment and precisely identify the cheating nodes without false accusations or stealing credits. The simulation results have also demonstrated that our routing protocols can improve the packet delivery ratio due to improving the routes' stability.

1.4 Outline of the Brief

The remainder of this brief is organized as follows. In Chap. 2, we present the system model including the network and communication models and the threat and trust models. In Chap. 3, we first explain the payment model that is specifically tailored to cooperation stimulation in MWNs. Then, we explain and evaluate our proposals ESIP and RACE. The trust/reputation system, TRIPO, and the routing protocols are presented and evaluated in Chap. 4. In this chapter, we will discuss how the payment reports can be processed to calculate the nodes' trust values and how the malicious nodes can be identified. We will also discuss how the routing protocols establish routes through the highly trusted nodes having sufficient energy. Finally, conclusion and future research directions will be discussed in Chap. 5.

References

1. X. J. Li, B. C. Seet, and P. H. J. Chong.   Multihop cellular networks: Technology and economics. *Computer Networks*, 52:1825–1837, 2008.
2. C. Gomes and J. Galtier.   Optimal and fair transmission rate allocation problem in multihop cellular networks.   *Lecture Notes in Computer Science, Springer Berlin/Heidelberg*, 5793:327–340, Aug. 2009.
3. H. Wu, C. Qios, S. De, and O. Tonguz.   Integrated cellular and ad hoc relaying systems: icar. *IEEE Journal of Selected Areas in Communications*, 19(10):2105–2115, Oct. 2001.
4. A. Abdrabou and W. Zhuang. Statistical qos routing for ieee 802.11 multihop ad hoc networks. *IEEE Transactions on Wireless Communications*, 8(3):1542–1552, Mar. 2009.
5. S. Bah and R. Glitho and R. Dssouli. Sip servlets for service provisioning in multihop cellular networks: High-level architectural alternatives. *Proc. of IEEE CCNC*, pages 127–131, 2008.
6. M. Kubisch, S. Mengesha, D. Hollos, H. Karl, and A. Wolisz. Applying ad-hoc relaying to improve capacity, energy efficiency, and immission in infrastructure-based wlans. *Proc. of Kommunikation in Verteilten Systemen, Leipzig, Germany*, pages 195–206, 2003.
7. R. Schoenen, R. Halfmann, and B. H. Walke. Mac performance of a 3gpp-lte multihop cellular network. *Proc. of IEEE ICC*, pages 4819– 4824, 2008.
8. P. Gupta and P. Kumar. The capacity of wireless networks. *IEEE Transactions on Information Theory*, 46(2):388–404, Mar. 2000.
9. C. Chou, D. Wei, C. Kuo, and K. Naik. An efficient anonymous communication protocol for peer-to-peer applications over mobile ad-hoc networks. *IEEE Journal on Selected Areas in Communications*, 25(1):192 – 203, Jan. 2007.
10. H. Gharavi. Multichannel mobile ad hoc links for multimedia communications. *Proc. of IEEE*, 96(1):77–96, Jan. 2008.
11. D. Johnson, D. Maltz, and Y. Hu.   The dynamic source routing protocol for mobile ad hoc networks (dsr). *technical report, IETF MANET Working Group*, Feb. 2007.
12. S. Marti, T. Giuli, K. Lai, and M. Baker. Mitigating routing misbehavior in mobile ad hoc networks. *Proc. of ACM International Conference on Mobile Computing and Networking (MobiCom'00), Boston, Massachusetts, USA*, pages 255–265, Aug. 6–11 2000.
13. P. Michiardi and R. Molva. Simulation-based analysis of security exposures in mobile ad hoc networks. *Proc. of European Wireless Conference, Florence, Italy*, Feb. 25–28 2002.
14. J. Hu. Cooperation in mobile ad hoc networks. *Technical report TR-050111, Computer Science Department, Florida State University, Tallahassee*, Jan. 2005.
15. G. Marias, P. Georgiadis, D. Flitzanis, and K. Mandalas. Cooperation enforcement schemes for manets: A survey. *Wiley's Journal of Wireless Communications and Mobile Computing*, 6(3):319–332, 2006.
16. Y. Zhang and Y. Fang. A fine-grained reputation system for reliable service selection in peer-to-peer networks. *IEEE Transactions on Parallel and Distributed Systems*, 18(8):1134–1145, Aug. 2007.
17. S. Lee, G. Pan, J-S Park, M. Gerla, and S. Lu.   Secure incentives for commercial ad dissemination in vehicular networks. *Proc. of MobiHoc'07*, Sep. 2007.
18. R. Lu, X. Lin, H. Zhu, C. Zhang, P.H. Ho, and X. Shen. A novel fair incentive protocol for mobile ad hoc networks. *Proc. of IEEE WCNC'08, Las Vegas, Nevada, USA*, March 31 - April 3 2008.
19. Y. Zhang and Y. Fang. A secure authentication and billing architecture for wireless mesh networks. *ACM Wireless Networks*, 13(5):569–582, Oct. 2007.
20. F. Milan, J. Jaramillo, and R. Srikant. Achieving cooperation in multi-hop wireless networks of selfish nodes. *Proc. of workshop on Game Theory for Communications and Networks, Pisa, Italy*, Oct. 14 2006.
21. L. Feeney. An energy-consumption model for performance analysis of routing protocols for mobile ad hoc networks. *Mobile Networks and Applications*, 3(6):239–249, 2001.

22. Y. Lin and Y. Hsu. Multihop cellular: A new architecture for wireless communications. *Proc. of IEEE INFOCOM'00*, 3:1273–1282, Mar. 26–30 2000.
23. J. Palmer and L. Eriksen. Digital newspapers explore marketing on the internet. *ACM Communications*, 42(9):33–40, 1999.
24. G. Iannaccone, C. Chuah, R. Mortier, S. Bhattacharyya, and C. Diot. Analysis of link failures in an ip backbone. *Proc. of IMW 2002, ACMPress, Marseille, France*, Nov. 2002.
25. X. Li, Z. Li, M. Stojmenovic, V. Narasimhan, and A. Nayak. Autoregressive trust management in wireless ad hoe networks. *Ad hoc and Sensor Wireless Networks*, 16:229–242, Feb. 2012.
26. G. Indirania and K. Selvakumara. A swarm-based efficient distributed intrusion detection system for mobile ad hoc networks (manet). *International Journal of Parallel, Emergent and Distributed Systems*, 2013.
27. H. Li and M. Singhal. Trust management in distributed systems. *IEEE Computers*, 40(2):45–53, Feb. 2007.

Chapter 2
System Model

2.1 Network and Communication Models

For military and disaster recovery applications, the multi-hop wireless network can be considered ephemeral because it is used for a specific purpose and short duration. This brief considers the civilian applications of MWNs, where the network has long lifetime and the nodes have long-term relations with the network. Thus, with every interaction, there is always an expectation of future reaction.

As illustrated in Fig. 2.1, the considered multi-hop wireless network has an off-line trusted party (Tp) and mobile nodes. The mobile nodes have different hardware and energy capabilities. We assume that the clocks of the nodes are synchronized. The details of this synchronization process are out of the scope of the brief, but several mechanisms have been proposed to synchronize the nodes' clocks [1]. We consider only the unicast communications. If a source node needs to communicate with a remote destination node, the mobile nodes should act as routers and relay the source node's packets to the destination node.

The mobile nodes should contact Tp periodically at least once during a time interval, called updating time, that can be in the range of few days. During this connection, the nodes submit the payment reports and the *Evidences* (if requested) and receive renewed certificates. Without holding a valid certificates, the nodes cannot work as source, destination, or intermediate nodes. During this connection, the nodes can also purchase credits with real money. This can enable the nodes that cannot earn sufficient credits, such as those at the network border, to communicate. This can also protect the network from credit decline because the total charges may be more than the rewards, as will be discussed later. The connection to Tp can occur via the cellular networks' base stations, Wi-Fi hot spots, or wired networks such as the Internet.

Tp and its public key are known for all the mobile nodes. Tp stores and manages the nodes' credit accounts and trust values. When Tp receives the payment reports of a session, it verifies them. If the reports are fair, Tp updates the involved nodes' credit accounts and trust values. For the cheating reports, it requests the *Evidences*

M.M.E.A. Mahmoud and X. Shen, *Security for Multi-hop Wireless Networks*,
SpringerBriefs in Computer Science, DOI 10.1007/978-3-319-04603-7_2,
© The Author(s) 2014

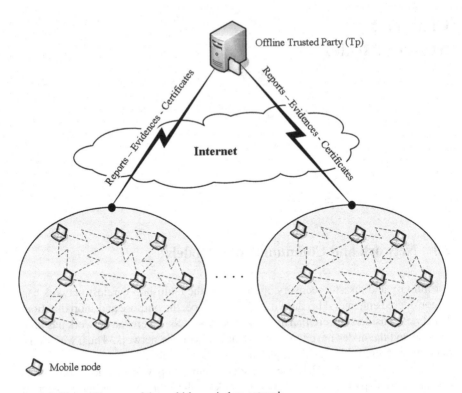

Offline Trusted Party (Tp)

Internet

Mobile node

Fig. 2.1 The architecture of the multi-hop wireless network

to identify the cheating nodes that send incorrect payment reports. Tp evicts the cheating nodes by denying renewing their certificates. It issues private/public key pair and a limited-time certificate with a unique identity for each node. For example, $\mathcal{N}_A$ will receive an identity ID_A and certificate $Cert_A$.

An identity-based key exchange protocol based on bilinear pairing is used by ESIP. The protocol is efficient because two nodes can compute a shared symmetric key without the need for exchanging messages. Tp generates a prime $\mathcal{P}$, a cyclic additive group $\mathcal{G}$, and a cyclic multiplicative group $\mathcal{G}_T$ of the same order $\mathcal{P}$ such that an efficiently computable bilinear pairing $\hat{e} : \mathcal{G} \times \mathcal{G} \longrightarrow \mathcal{G}_T$ is known. The bilinear mapping has the following properties:

- **Bilinear**: $\hat{e}(aP, bQ) = \hat{e}(bP, aQ) = \hat{e}(P, Q)^{ab} \forall P, Q \in \mathcal{G}$ and $a, b \in \mathcal{Z}_P^*$.
- **Non-degeneracy**: $\hat{e}(P, Q) \neq 1_{\mathcal{G}_T}$.
- **Symmetric**: $\hat{e}(P, Q) = \hat{e}(Q, P) \forall P, Q \in \mathcal{G}$.
- **Admissible**: There is an efficient algorithm that can compute $\hat{e}(P, Q) \forall P, Q \in \mathcal{G}$.

The bilinear mapping $\hat{e}$ can be implemented efficiently using the Weil and Tate pairings on elliptic curves [2]. Tp selects a random element $\mu \in \mathcal{Z}_P^*$ known as the master key, and computes the secret keys for the nodes based on their identities. The secret key for node ID_i is $Sk_i = \mu \cdot \mathcal{H}(ID_i) \in \mathcal{G}$, where $\mathcal{H} : \{0, 1\}^* \rightarrow \mathcal{G}$.

Each node, e.g., $\mathcal{N}_A$ has to register with Tp to receive a unique identity (ID_A), symmetric key K_A, private/public key pair, a valid certificate, and the required cryptographic data to enable any two nodes to share a symmetric key. The symmetric key is used to submit the payment reports and the private/public keys are required to act as source, intermediate or destination node.

2.2 Threat and Trust Models

The mobile nodes are probable attackers but Tp is fully secure. The mobile nodes are autonomous and self-interested and thus motivated to misbehave to maximize their welfare and minimize their contributions. However, Tp is run by an operator that is interested in ensuring the network secure operation. As proven in [3], it is impossible to realize secure payment between two entities without involving a trusted third party.

The attackers have full control on their devices, and thus they can change their normal operation and obtain the cryptographic credentials. These strong assumptions are necessary due to using payment in the network. Non-participation in packet relay is not an abuse because the nodes are stimulated and not forced to relay others' packets with their own devices, but the large rate of packet dropping is an abuse due to disrupting the packets routing process.

Some attackers may act rationally by misbehaving only when they can achieve more benefits than behaving honestly. For example, they may attempt to attack the payment system to steal credits, pay less, and communicate freely. Some adversaries may report incorrect battery energy level to increase their chance to be selected by the routing protocol, e.g., to earn more credits. On the contrary, other attackers may act irrationally without considering their interests. For example, they may launch *Denial-of-Service* attacks by involving their devices in communication routes and dropping the data packets intentionally to break the routes. When an attacker $\mathcal{N}_B$ receives packets from $\mathcal{N}_A$ to forward to the next node in the route, $\mathcal{N}_B$ drops the packets and keeps silent to let $\mathcal{N}_A$ believe that $\mathcal{N}_B$ is out of transmission range and the link between them is broken. The attackers may launch *Black-Hole* attack by continuously breaking all the routes they participate in. They may also launch *Gray-Hole* attack by intentionally breaking some routes and behaving regularly in other routes to circumvent Tp, but the ratio of the broken routes should be large to launch effective attacks. These attacks may be launched by compromised, malfunctioned, or low-resource nodes.

The adversaries may attempt to attack the trust system by launching *Trust-Boost* attack to falsely augment their trust and reputation values to escape the consequence of dropping packets or increase their chance to be selected in routes. They may also try to launch *False-Accusation* attacks to degrade honest nodes' trust values to evict them from the network.

The attackers may work individually or collude with each other to launch sophisticated attacks. The gained experience from the currently used protocols in civilian applications confirms that large-scale irrational collusion attacks are highly unlikely [4]. The trust/reputation systems are susceptible to the large-scale collusion attacks due to the nature of these systems. Our objective is to protect the payment against all types of collusion attacks, and protect the trust/reputation system against small-scale irrational collusion attacks and improve the system's robustness against large-scale attacks.

For the trust models, the nodes fully trust Tp to perform billing and auditing and trust calculations, but Tp does not trust any node in the network.

References

1. B. Wehbi, A. Laouiti, and A. Cavalli. Efficient time synchronization mechanism for wireless multi hop networks. *Proc. IEEE Personal, Indoor and Mobile Radio Comm. (PIMRC)*, 2008.
2. D. Boneh and M. Franklin. Identity based encryption from the weil pairing. *Proc. of Crypto'01, LNCS, Springer-Verlag*, 2139:213–229, 2001.
3. H. Pagnia and F. Gartner. On the impossibility of fair exchange without a trusted third party. *Technical Report TUD-BS-1999-02, Darmstadt University of Technology*, Mar. 1999.
4. C. Cachin, K. Kursawe, A. Lysyanskaya, and R. Strobl. Asynchronous verifiable secret sharing and proactive cryptosystems. *Proc. ACM Conference on Computer and Communications Security, CCS02*, pages 88–97, 2002.

Chapter 3
Efficient Incentive Scheme

Figure 3.1 shows the architecture of the proposed security protocols/schemes. In the *Communication* phase, the nodes establish routes and transmit data without involving a trusted party or a central unit. In *Route Establishment* phase, the nodes use either *BAR (Best Available Route)* or *SRR (Shortest Reliable Route)* routing protocol to establish stable and reliable routes. These routing protocols will be explained in details in Sect. 4.3. In *Data Transmission* phase, the source node transmits messages to the destination. The source, intermediate, and destination nodes store security tokens called *Evidences* and compose payment reports. The nodes use a communication protocol, called ESIP, that can secure the payment with limited use of public-key cryptography. This protocol will be explained in Sect. 3.2. The nodes submit payment reports containing the payment data of different sessions to Tp to redeem the payment. In *Classifier* phase, Tp classifies the reports into fair or cheating. For cheating reports, some nodes in a session submit incorrect payment reports, e.g., to steal credits. These reports are passed to *Cheaters Identification* phase that requests the *Evidences* to identify the cheating nodes. In *Processing* phase, the payment reports are processed to extract contextual and financial information. The financial information is passed to the *Credit-Account Update* phase to update the nodes' credit accounts. In Sect. 3.1, we will discuss the payment model that is used to update the nodes' credit accounts. In *Trust Update* phase, the contextual information is passed to the trust system to update the nodes' trust and reputation values. The malicious and cheating are evicted by denying renewing their certificates. In Sect. 3.3, we will explain RACE, an efficient scheme to submit/process payment reports. In Sects. 4.1 and 4.2, we will explain the trust system and TRIPO, a mechanism to identify the packet dropping attackers.

M.M.E.A. Mahmoud and X. Shen, *Security for Multi-hop Wireless Networks*,
SpringerBriefs in Computer Science, DOI 10.1007/978-3-319-04603-7_3,
© The Author(s) 2014

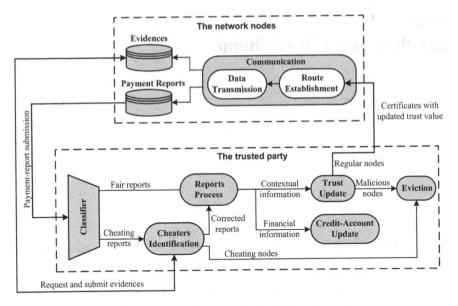

Fig. 3.1 The architecture of the proposed security protocols/schemes

3.1 Payment Model

Micropayment schemes [1, 2] are electronic payment schemes for frequent and low-value payments. The schemes were originally designed for the Internet electronic commerce applications to take advantage of the high volume of viewers by offering content for low price. Examples of these applications include buying data or news, listening to a song, playing an online game, and reading an article in a journal [3]. In order to effectively implement micropayment in MWNs, the payment model should consider the differences between Web-based applications and cooperation stimulation.

These differences are summarized in Table 3.1. For Web-based applications, a transaction usually has one customer and one merchant, and the merchants' number is low and their identities are known before the transaction is held. For cooperation stimulation, each transaction usually has two customers (the source and destination nodes) and multiple merchants (the intermediate nodes), the merchants' number is large because any node can work as a merchant (or packet relay), and the merchants' identities are known only at the transaction (session) time due to the nodes' mobility. Moreover, the relation between a customer and a merchant is usually short due to the network's dynamic topology. The nodes are involved in low-value transactions very often because once a route is broken, which is often due to the nodes' mobility and the channel impairment, a new transaction should be held to re-establish the route. In MWNs, the nodes have low resources such as energy and storage area comparing to the computers' large resources in Web-based applications.

Table 3.1 Properties of Web-based applications and cooperation stimulation

		Web-based applications	Cooperation stimulation
P1	Transactions' parties	One customer and one merchant	One or more merchants and two customers
P2	Merchants' number	Low	Large
	Merchants' identities	Known in advance	Unknown in advance
P3	Customer-merchant relation	Long	Very short
P4	Transaction frequency	High	Very high
P5	Transaction value	Low	Very low
P6	Easiness of misbehavior	Very easy	Less
P7	Nodes' resources	High	Low

Although security is important in all payment applications, the attacks can be launched easier in Web-based applications because it is easier to launch attacks across the Internet than tampering devices.

3.1.1   Parties and Relations

The payment model has three basic parties: the *customer* or the source and destination nodes; the *merchant* or the intermediate nodes; and the *bank* or Tp. Figure 3.2 portrays the relations among the different parties in our payment model. The interactions between these parties can be divided into three phases: *Certificate Issuing*; *Payment*; and *Redemption*. In *Certificate Issuing* phase, a customer has to register with the bank to create an account, and the bank issues a short-lifetime certificate, e.g., for seven to ten days. The customer contacts the bank periodically to renew his certificate and pay for the services (packet relay) he received from the merchants. In *Payment* phase, the customer's certificate enables him to issue digital receipts (undeniable payment proofs) to transact with merchants without involving the bank, i.e., customers can mine their own electronic coins without the need for direct verification by the bank. In *Redemption* phase, each merchant claims its payment by submitting payment receipts. Tp verifies the receipts and clears them by rewarding the merchants and charging the customers. This payment system architecture has two important properties that can make using micropayment in multi-hop wireless networks effective: *no need for tamper-proof-device* and *flexibility*.

3.1.1.1   No Need for Tamper-Proof-Device

The existing incentive schemes can be classified into two categories: central-bank based and tamper-proof device (TPD) based. In TPD-based incentive schemes [4–7], a tamper proof device is installed in each device to manage its credit account

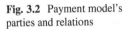

Fig. 3.2 Payment model's parties and relations

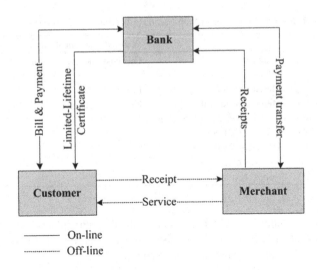

and secure its operation. The self-generated and forwarded packets are passed to the node's tamper proof device to decrease and increase its credit account, respectively. In central-bank-based incentive schemes [8–11], a centralized unit stores and manages the nodes' accounts.

The TPD-based incentive schemes may not find widespread acceptance for the following reasons. First, the assumption that TPDs cannot be tampered is neither secure nor practical for MWNs. That is because the nodes are autonomous and self-interested and the attackers can communicate freely in an undetectable way if they could compromise the TPDs. Moreover, since the security protection of these protocols completely fails if the TPDs are tampered, only a small number of manufactures can be trusted to make the network nodes, which is too restrictive for civilian networks. Second, a node cannot communicate if it does not have sufficient credits at the communication time. Unfortunately, the nodes at the network edge cannot earn as many credits as the nodes at other locations because they are less frequently selected by the routing protocol. Furthermore, the credit distribution has direct impact on the communication activity, e.g., if a small number of nodes have large ratio of the network credits, the communication activity significantly degrades because the rich nodes are not motivated to cooperate and the poor nodes cannot initiate communications. Finally, since credits are cleared in real-time, the communication activity degrades if the network does not have enough credits circulating around. In [7], it is shown that the overall credits in the network gradually decline because the total charges are not necessarily equal to the total rewards. This is because the source node is fully charged after sending each packet, but some intermediate nodes may not be rewarded when the route is broken. In [12], a compensation mechanism is used to change the packet-relaying price proportionally to the nodes' speed to avoid the credit decline. However, the compensation mechanism has to avoid credit inflation and depletion. For credit inflation, the nodes become rich and thus less motivated to relay others' packets, whereas for credit depletion, the nodes become poor and incapable of initiating communication. The design of a decentralized compensation mechanism to stabilize the amount of credits in the network is difficult especially in large-scale networks.

In our payment model, Tp sells credits for real money and converts credits to real money. This can motivate the rich nodes, who have many credits, to cooperate, improve credit distribution, and protect the network from credit decline. This can also enable the nodes that cannot earn enough credits because they are less frequently selected by the routing protocol (such as those at the network border) to communicate. However, we do not consider this a fairness problem because the philosophy behind the incentive schemes is that packet relay is a service and not an obligation. This service may not be requested from some nodes, i.e., the customers request the packet-relay service from the best service providers. If the routing protocol routes packets to the border nodes to enable them to earn credits, obviously, the network performance will degrade because the routes may be long. Due to the nodes' mobility, the border nodes can change their location and earn more credits as shown in [12]. Moreover, since the border nodes do not relay as many packets as others, it is fair to charge them real money to compensate the other nodes that relayed more packets.

3.1.1.2 Flexibility

There are two ways for managing the electronic payment: *on-line* and *off-line*. For on-line payment, a merchant verifies the payment with the bank before serving the customer; and for off-line payment, a merchant serves the customer without involving the bank at the transaction time, i.e., instead of interacting with the bank in each transaction, the merchants accumulate the payments and redeem them in batch later. The payment management can also be classified into *credit* (or postpaid) and *debit* (or pre-paid). For credit payment systems, the customers are served first and charged later, e.g., the customers issue receipts to the merchants that submit them to the bank to redeem the payment. A customer's account will not be debited until the payment reports are processed. For debit payment systems, the customers' accounts are charged before they are served, e.g., customers buy electronic coins in advance from the bank to pay the merchants, or the bank has to be interactively involved in each session.

Off-line and credit payment systems are better for the effective use of micropayment in MWNs for the following reasons. First, the connection with the bank may not be available on a regular basis, and even if it is available, involving a centralized unit in each transaction is inefficient and creates bottleneck at the bank due to the high frequency of low-value transactions (P4 and P5 in Table 3.1). Second, customers generate their own coins (or receipts), which offer flexibility to the system. Coins are generated on-demand and customers do not need to frequently contact the bank to buy coins. In [7], it is shown that some nodes cannot communicate because they cannot find a service point to get credits. Moreover, generating coins to pay for a specific merchant [13] is not practical due to the large number of potential merchants in the network, and generating general coins to pay for any merchant is vulnerable to *Double-Spending* attack or requires interactive and frequent contact with the bank. In *Double-Spending* attack, the attackers spend the same coin multiple times because it is difficult to know if a coin has been used before or not.

Although, the developed payment architecture has many positives, it is obvious that reducing the overhead of submitting and clearing the payment receipts is essential for the effective implementation for the following reasons. First, since the transactions' number is expected to be large and multiple merchants may be involved in a transaction (P1 and P2 in Table 3.1), generating a receipt per message [8] or per customer [10] significantly increases the receipts' number, and thus the transaction value may not cover its processing cost (P5 in Table 3.1). Second, the nodes have low resources (P7 in Table 3.1) so the overhead of storing and submitting a large number of receipts may stimulate the nodes to behave selfishly. What may make this worse is that the nodes need to store the receipts for some time until a connection to the bank becomes available.

3.1.2 Charging and Rewarding Policy

In most existing incentive schemes [4, 8–10, 12], only the source node is charged no matter how the destination node benefits from the communication. We argue that a fair charging policy is to support cost sharing between the source and destination nodes when both of them benefit from the communication [14]. The payment-splitting ratio is adjustable and service-dependent, e.g., a domain name system (DNS) server should not pay for name resolution. The source and destination nodes agree on the payment-splitting ratio during the session establishment phase. For the rewarding policy, some incentive schemes, such as [15, 16], use variable packet relaying rewards that correspond to the incurred energy in relaying the packets. This rewarding policy is difficult to be implemented in practice without involving complicated route discovery process and calculation of en-route individual payments. Therefore, similar to [8–10, 13, 17–22], we use fixed rewarding rate, e.g., λ credits per unit-sized message.

In MWNs, packet loss may occur for non-malicious reasons due to node mobility, packet collision, channel impairment, etc. Ideally, any node that has ever tried to forward a packet should be rewarded no matter whether the packet eventually reaches its destination or not because forwarding a packet consumes the node's resources. However, it is difficult to corroborate an intermediate node's packet forwarding in a trustable and distributed manner without involving too complicated design. For example, rewarding the nodes for relaying the route establishment packets or packet retransmissions complicates the incentive scheme and significantly increases the receipts' number because a large number of nodes may be involved in relaying route establishment packets and packet retransmissions may happen frequently in wireless networks. Our charging and rewarding policy rewards the intermediate nodes only for the delivered data packets, but the source and the destination nodes are charged for every transmitted message even if it does not reach the destination nodes as illustrated in Fig. 3.3a,b. For fair rewarding policy, the value of λ is determined to compensate the nodes for the consumed resources in relaying route establishment packets, packet retransmissions, and undelivered packets. In Sect. 3.4, we will argue that this charging and rewarding policy can discourage misbehavior and encourage relaying packets faithfully.

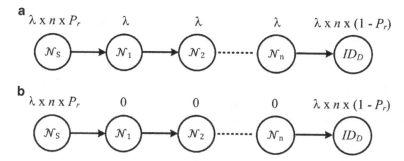

Fig. 3.3 The payment rewarding and charging policy. (**a**) Delivered packets. (**b**) Undelivered packets

3.2 ESIP: Communication Protocol with Limited Use of Public Key Cryptography

In this section, we present ESIP, a communication protocol that can secure the payment with limited use of public-key cryptography [23, 24]. The source node transmits data packets to the destination node that replies with *ACK* packets to acknowledge receiving them. As illustrated in Fig. 3.4, each node in the route should share a symmetric key with the source node. This key will be used to compute the messages' keyed hash values. Two nodes with identity and secret key pairs (ID_S, SK_S) and (ID_A, SK_A) can independently compute the shared key as follows:

$$
\begin{aligned}
K_{SA} &= \hat{e}(H(ID_A), Sk_S) \\
&= \hat{e}(H(ID_A), \mu \cdot H(ID_S)) \\
&= \hat{e}(\mu \cdot H(ID_A), H(ID_S)) \qquad \text{(Bilinear property)} \\
&= \hat{e}(Sk_A, H(ID_S)) \\
&= \hat{e}(H(ID_S), Sk_A) \qquad \text{(Symmetric property)} \\
&= K_{AS}
\end{aligned}
$$

As shown in Fig. 3.5, the source and destination nodes generate hash chains by iteratively hashing random values. The source node hashes V_S^N N times to obtain the final hash value V_S^1 and the destination node hashes V_D^N $N + 1$ times to obtain final hash value V_D^0, where $V_S^{i-1} = H(V_S^i)$ and $V_D^{i-1} = H(V_D^i)$. The source node releases one hash value as a proof of sending a packet and the destination node releases one hash value as a proof for receiving a packet. The hash values are released in the direction from V_S^1 to V_S^N and V_D^0 to V_D^N. Payment non-repudiation can be achieved because it is infeasible to compute V_S^i from V_S^{i-1} or V_D^i from V_D^{i-1} for $2 \leq i \leq N$. In order to authenticate the hash chains and link them to the route, the source and destination nodes sign the roots of the hash chains and other information such as the identities of the nodes (the payers and the payees) in the route (e.g., $R = ID_S, ID_A, ID_B, ID_C, ID_D$ in Fig. 3.4), the route establishment time stamp (Ts), the payment ratio (Pr), etc. The source node's signature is sent in the first data packet while the destination node's signature is sent in the *RREP* packet. More details about these signatures will be given when we discuss the routing protocols in Sect. 4.3.

3.2.1 Data Generation and Relay

For the first data packet in a series of at most N packets, Fig. 3.6 shows that the source node appends the message M_1, V_S^1 and its signature $Sig_S(V_S^1, ..)$. This signature can prove the source node's approval to pay for relaying its packets, i.e., the sender cannot deny generating the hash chain or initiating the session. In order

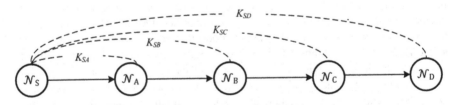

Fig. 3.4 The source node shares a key with each node in the route

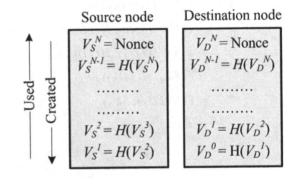

Fig. 3.5 The source and destination nodes' hash chains

to ensure the message's authenticity and integrity at each intermediate node, the message's hash value $(H(M_1))$ can be included in the signature, but this increases the *Evidence* size because $H(M_1)$ has to be attached to the *Evidence*. Instead, the source node attaches the hash series $HS(M_1)$ which has a truncated keyed hash value for each node, e.g., $HS(M_1) = H_{K_{SA}}(M_1), H_{K_{SB}}(M_1), H_{K_{SC}}(M_1), H_{K_{SD}}(M_1)$ in Fig. 3.6. Each intermediate node verifies the source node's signature to ensure that it will compose a valid *Evidence* and thus it will be rewarded for relaying the packets. It also verifies its message's truncated hash value to ensure the message's authenticity and integrity, and then relays the packet after dropping its hash value as shown in Fig. 3.7. Each intermediate node stores the source node's signature and V_S^1 to be used in composing the *Evidence*.

For data packet number X and $X > 1$, Fig. 3.6 illustrates that the source node appends the pre-image of the last released hash value (V_S^X) as an approval to pay for one more packet. It also attaches the message's hash series $(HS(M_X))$. Before relaying the packet, each intermediate node verifies its truncated keyed hash value, verifies that V_S^{X-1} is generated from hashing V_S^X, and relays the packet after dropping its keyed hash value. The intermediate nodes store only the last hash chain element to be used in composing the *Evidence*, i.e., after receiving the Xth data packet, the intermediate node deletes V_S^{X-1} and store V_S^X. V_S^X alongside the source node's signature are sufficient to prove that X messages have been transmitted. Each node in the route restarts a timer each time the node relays a packet. The route is considered broken when the timer expires.

3.2.2 ACK Generation and Relay

Upon receiving the Xth data packet and $X \leq N$, Fig. 3.6 shows that the destination node sends back *ACK* packet containing the pre-image of the last released hash value, or V_D^X, to acknowledge receiving the message in an undeniable way. Each intermediate node verifies that V_D^{X-1} is generated from hashing V_D^X, and stores the last hash value (V_D^X) to be used in composing the *Evidence*. Therefore, instead of generating a signature per *ACK* packet, one signature is generated per N ACKs. Payment non-repudiation and non-manipulation are achievable because the hash function is one-way, i.e., only the destination node could have generated the hash chain because it is infeasible to compute V_D^{X+1} from V_D^X.

As illustrated in Fig. 3.6, after releasing all the hash values of the first hash chain, the source and destination nodes create new hash chains by iteratively hashing random values $V_S^N(1)$ and $V_D^N(1)$ N times. In the data packet number $N + 1$, the source node authenticates its new hash chain and links it to the route by signing the root of the new hash chain alongside other information. The roots of all used hash chains $(V_S^1$ and $V_S^1(1))$ are signed instead of signing only the last hash chain's root. By this way, the intermediate nodes store only the last signature for composing the *Evidence* because it can authenticate all the used hash chains. Similarly, in the *ACK* of the message number $N + 1$, the destination node sends its signature for

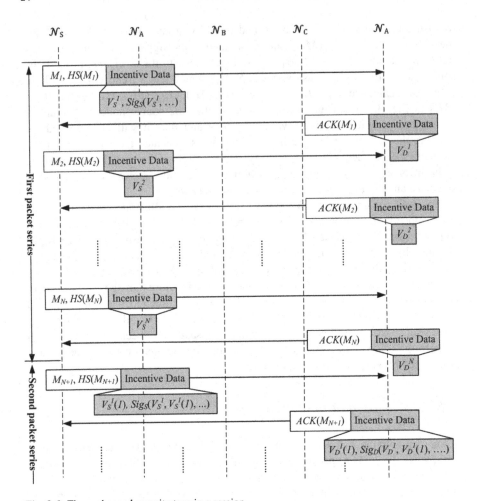

Fig. 3.6 The exchanged security tags in a session

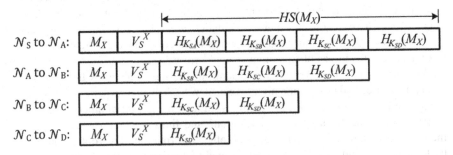

Fig. 3.7 The hop-by-hop security packet-overhead in the Xth data packet and $X > 1$

the roots of all used hash chains (V_D^0 and $V_D^1(1)$) alongside other information. More details about the information signed with the hash chains' roots will be given in Sect. 4.3. The nodes store the hash chains' roots and the last signature generated by the destination node for composing the *Evidence*.

3.3 RACE: Efficient Scheme for Submitting/Processing the Payment Data

In this section, we present RACE, a scheme to efficiently submit and process the payment data to Tp [25, 26].

3.3.1 Evidence Composition and Payment-Report Submission

Each node in a route should compose a payment report and a security *Evidence*. A payment report has the session identifier (SI), a flag bit (F), and the number of messages (X). SI has the identities of the nodes in the route including the source and destination nodes and the route establishment time stamp (Ts). The flag bit indicates whether the last received packet is data or *ACK*. It is zero if the last received packet is data and one if it is *ACK*. Since a connection to Tp may not be available on a regular basis, the nodes accumulate the payment reports and submit them in batch to Tp for redemption. Table 3.2 gives numerical example for the payment reports of $\mathcal{N}_A$. For the first report, $\mathcal{N}_A$ is the source node, $\mathcal{N}_C$ is the destination, and $\mathcal{N}_B$ and $\mathcal{N}_N$ are the intermediate nodes. In the report, $\mathcal{N}_A$ claims sending 12 messages but it did not receive the *ACK* of M_{12} because F is zero. For the second report, $\mathcal{N}_A$ is the destination node and claims receiving 17 messages and sending the *ACK* of M_{17}. For the third report, $\mathcal{N}_A$ is an intermediate node and claims receiving 15 messages to relay, but it did not receive the *ACK* of M_{15}.

Evidence is defined as information that is used to establish proof about the occurrence of an event or action, the time of occurrence, the parties involved in the event, and the result of the event. The purpose of an *Evidence* is to resolve a dispute about the amount of the payment. The general format of an *Evidence* is given in Fig. 3.8a, where the brackets "[Y]" mean that Y may not exist in some cases. An *Evidence* has two main parts called data (D) and security token (S_T). The data has SI, the roots and seeds of the source and the destination nodes' hash chains, and the last releases hash values ($V_D^L(v)$, $V_S^L(v)$). It also has the energy and trust requirements (E_r and T_r), and the maximum number of intermediate nodes (H_{max}).

Note that T_r will not be needed in case of *BAR* routing protocol. S_T is an undeniable cryptographic proof that prevents payment repudiation and manipulation, and thus ensures that the payment is undeniable, unmodifiable, and unforgeable. In order to significantly reduce the *Evidence*'s size, S_T is composed by hashing

Table 3.2 Numerical example for the payment reports of $\mathcal{N}_A$

Session Identifier	Messages' number (X)
$ID_A, ID_B, ID_N, ID_C, Pr_1, Ts_1, 0$	12
$ID_F, ID_C, ID_B, ID_H, ID_A, Pr_2, Ts_2, 1$	17
$ID_C, ID_F, ID_A, ID_H, Pr_3, Ts_3, 0$	15

a

Evidence

$D = SI, V_D^0, [V_D^1(1),...], V_S^1, [V_S^1(1),...], [V_D^N, V_S^N, ...], [V_D^L(v), V_S^L(v)], H_{max}, Er, Tr$

$S_T = H(\ \underline{Sig_S(SI, V_S^1, V_S^1(1) ...)}, \underline{Sig_D(Auth\_Code, V_D^0, V_D^1(1)....)}\ )$

b

Evidence(X)

$D = SI, V_D^0, V_S^1, [V_D^{X-1}], V_S^X, H_{max}, Er, Tr$

$S_T = H(\ \underline{Sig_S(SI, V_S^1)}, \underline{Sig_D(Auth\_Code, V_D^0)}\ )$

c

Evidence(X)

$D = SI, V_D^0, V_D^N, V_S^1, V_S^N, V_D^1(1), V_S^1(1), V_D^{X-1}(1), V_S^X(1), H_{max}, Er, Tr$

$S_T = H(\ \underline{Sig_S(SI, V_S^1, V_S^1(1))}, \underline{Sig_D(Auth\_Code, V_D^0, V_D^1(1))}\ )$

d

Evidence(X)

$D = SI, V_D^0, V_S^1, V_D^X, V_S^X, H_{max}, Er, Tr$

$S_T = H(\ \underline{Sig_S(SI, V_S^1)}, \underline{Sig_D(Auth\_Code, V_D^0)}\ )$

e

Evidence(0)

$D = SI, V_D^0, H_{max}, Er, Tr$

$S_T = H(\ \underline{Sig_D(Auth\_Code, V_D^0)}\ )$

Fig. 3.8 The formats of the payment *Evidence*s. (**a**) General *Evidence* format. (**b**) Last received packet is data and $1 \leq X \leq N$. (**c**) Last received packet is data and $N < X \leq 2N$. (**d**) Last received packet is *ACK* and $1 \leq X \leq N$. (**e**) The last received packet is *RREP*

the source and the destination nodes' last signatures instead of attaching the large-size signatures. The authentication code ($Auth\_Code$) has the source and intermediate nodes' signatures to hold them accountable for any misbehavior. In Sect. 4.3, more details will be given about $Auth\_Code$, E_r, T_r, and H_{max}. The *Evidence* size depends on the number of used hash chains because two hash values should be attached for each hash chain, and thus properly choosing the hash chain size can minimize the *Evidence* size.

Figure 3.8b shows the format of the *Evidence* when the last received packet is the Xth data packet and $1 \leq X \leq N$, i.e., only one hash chain is used in the route. If the route is broken after receiving the first data packet ($X = 1$), the *Evidence* does not have V_D^1 because the *ACK* packet is not received. For $1 < X \leq N$, the last hash value received from the destination node (V_D^{X-1}) is attached to the *Evidence*. Since the route is broken before receiving the *ACK* of message X, the last released hash value from the destination node is V_D^{X-1} and not V_D^X. Figure 3.8c gives the *Evidence* when a route is broken after receiving the Xth data packet and $N < X \leq 2N$. In this case, two hash chains are used. It can be seen that the *Evidence* has the seed and the root of the first hash chain, the root of the second hash chain, and the last released hash value ($V_D^{X-1}(1)$ and $V_S^X(1)$). Moreover, the two hash chains' roots are included in the source and the destination nodes' signatures. Figure 3.8d gives the *Evidence* when the last received packet is the *ACK* of the Xth message. Note that an *Evidence* has V_S^X when the node receives the Xth data packet and V_D^X when it receives the Xth *ACK* packet. V_S^X is an undeniable proof for sending X packets by the source node and V_D^X is an undeniable proof for receiving X packets by the destination node. Figure 3.8e shows the format of the *Evidence* when the last received packet is *RREP*. The *Evidence* does not have the source node's packet because it is sent in the first data packet.

Evidences have the following main features:

1. *Evidences* are unmodifiable: If X messages are delivered, the intermediate nodes can compose *Evidences* for fewer than X messages, but not for more. This is because the intermediate nodes have the necessary hash chain elements and signatures for composing *Evidences* for fewer than X messages. However, the intermediate nodes cannot compose *Evidences* for more than X because it is computationally infeasible to forge the source and destination nodes' signatures or compute hash chain elements that are not released.

2. The source and destination nodes of a legitimate session can collude to create *Evidences* for any number of messages because they can compute the necessary hash chain elements and signatures.

3. *Evidences* are unforgeable: without collusion, the nodes cannot create *Evidences* for sessions that did not happen because the security token has the nodes' signatures and it is infeasible to forge the signatures. Note that the *Auth_Code* has the intermediate nodes' signatures.

4. *Evidences* are undeniable: This is necessary to enable Tp to verify them and secure the payment. A source node cannot deny initiating a session or the amount of payment because its signature is included in the *Evidence* and it is infeasible to compute the hash chain elements.

5. An honest intermediate node can always compose valid *Evidence* even if the route is broken or the other nodes in the route collude to manipulate the payment of the node. This is because it can verify the *Evidences* to avoid being fooled by the attackers.

***Evidence* Aggregation Technique:** Reducing the storage area of the *Evidences* is important because they should be stored until the Tp receives the payment reports from all the nodes in the route and clears the payment. Onion hashing technique can be used to aggregate *Evidences*. The underlying idea is that instead of storing one *Evidence* per session, one aggregated *Evidence* can be computed to prove the credibility of the payment of a group of sessions. In Fig. 3.9, $D(i)$ and $S_T(i)$ are the data and the security token of *Evidence* number i, respectively. The figure shows that the aggregated *Evidence* has the concatenation of the individual *Evidences'* data and one aggregated security token that is computed by onion hashing the security tokens of the individual *Evidences*. $S_T(1)$ and $S_T(2)$ are concatenated and hashed together, and then $S_T(3)$ is added to the aggregated security token by adding one hashing layer and so on. The onion-hashing technique enables the nodes to aggregate a recent *Evidence* to the aggregated *Evidence*, i.e., the *Evidences* are always stored in aggregated format to reduce their storage area. The aggregated security token has the same size of those of the individual *Evidences*, but it can prove the credibility of the payment of multiple sessions. The technique is called onion hashing because each aggregation operation requires adding one hashing layer.

However, the *Evidence* aggregation process is irreversible because the hash function is unidirectional, i.e., the aggregated *Evidence* cannot be decomposed to individual *Evidences*. Thus, if Tp requests an *Evidence* that is aggregated in the aggregated *Evidence*, the node has to submit the aggregated *Evidence* and Tp has to verify all the security tokens of the sessions of the aggregated *Evidence*, instead of verifying only the security token of the requested *Evidence*. For example, if *Evidences* E_1, E_2, and E_3 are aggregated to E_A and Tp requests E_2, the node has to submit E_A and Tp has to verify the node's credibility in the three *Evidences*.

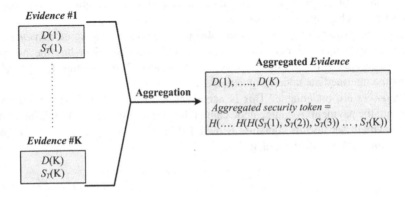

Fig. 3.9 Aggregated *Evidence*

Therefore, *aggregating more Evidences can further reduce their storage area, but with more communication and processing overhead if an Evidence is requested.* This is acceptable because *Evidences* are requested only in case of cheating and as will be explained in Sect. 3.3.3, *Evidences* are requested only from few nodes instead of all the nodes in the cheating reports. The *Evidence* aggregation level can be flexible and dependent on the available memory space, e.g., a storage-constrained node can aggregate all *Evidences* in only one aggregated *Evidence*.

Reports and *Evidences* submission: As shown in Fig. 3.10, $\mathcal{N}_A$ sends a *Report Submission Packet* (*RSP*) to Tp at time t_n to submit the reports of the routes established since the last contact at t_{n-1}. The packet has the reports ($Reports[t_{i-1}, t_i]$), time stamp (Ts), and a keyed hash value ($H_{KA}()$) to ensure the packet's integrity and authenticity, where K_A is the long-term symmetric key shared between $\mathcal{N}_A$ and Tp. Thus, Tp can make sure that the packet has not been manipulated and the reports are indeed sent by the intended node, which is important to secure the payment and hold the nodes accountable for any misbehavior. If Tp requests *Evidences* from $\mathcal{N}_A$, it sends an *Evidences Request Packet* (*EREQ*) having the session identifiers of the reports that their *Evidences* are requested ($Ses\_IDs[t_{n-2}, t_{n-1}]$). $\mathcal{N}_A$ replies with *Evidences Reply Packet* (*EREP*) having the requested *Evidences* ($Req\_Evs[t_{n-2}, t_{n-1}]$). If $\mathcal{N}_A$ is honest, Tp sends a *Renewed Certificate Packet* (*RCP*) having a renewed certificate for $\mathcal{N}_A$ with the same identity and public/private keys, but with updated lifetime and trust values. Therefore, only the efficient hashing operations are used to submit the reports and *Evidences* securely to Tp. Note that

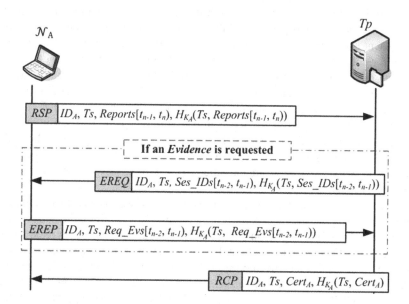

Fig. 3.10 The submission of reports and *Evidences*

Table 3.3 Numerical examples for fair payment reports

Case number		$\mathcal{N}_S$	$\mathcal{N}_A$	$\mathcal{N}_B$	$\mathcal{N}_C$	$\mathcal{N}_D$
1	X	11	11	11	11	11
	F	1	1	1	1	1
2	X	11	11	11	11	11
	F	0	0	1	1	1
3	X	8	8	7	7	7
	F	0	0	1	1	1
4	X	–	–	–	0	0
	F	–	–	–	0	0
5	X	1	1	1	0	0
	F	0	0	0	0	0

RSP and RCP packets are also required in receipt-based incentive schemes to submit the receipts.

3.3.2 Reports Classification

The network nodes periodically submit payment reports to Tp to redeem the payment. As shown in Fig. 3.1, after receiving a route's payment reports, Tp verifies them by investigating the consistency of the reports, and classifies them into fair or cheating. Tp uses the report's unique identifier (SI) to link all the reports of one route. For fair reports, the nodes submit correct payment reports, but for cheating reports, at least one node does not submit the reports or submits incorrect reports, e.g., to steal credits or pay less. For the cheating reports, the nodes' reports are inconsistent and cannot occur without cheating from at least one node. Fair reports can be for complete or broken routes. For complete route reports, all the nodes in the route report the same number of messages and F of one. If a route is broken during relaying the Xth data packet, the reports of the nodes from $\mathcal{N}_S$ to the last node that received the packet submit X for the messages' number and F of zero, but the other nodes submit $X - 1$ and F of one. If a route is broken during relaying the Xth ACK packet, all the nodes in the route submit X messages, and the nodes from $\mathcal{N}_D$ to the last node that received the ACK submit F of one but the other nodes submit F of zero. The reports are classified as cheating if they do not apply to one of the aforementioned rules.

Table 3.3 gives numerical examples for fair reports. Case 1 is reports for complete route and Cases 2 to 5 are reports for broken routes. For Case 1, all the nodes report the same number of messages and F of one. For Case 2, the route was broken during relaying the ACK packet number 11 and $\mathcal{N}_B$ is the last node that received the packet. For Case 3, the route was broken during relaying the data packet number 8 and $\mathcal{N}_A$ is the last node that received the packet. For Case 4, the route was broken during relaying the $RREP$ packet and $\mathcal{N}_C$ is the last node that received the packet. $\mathcal{N}_B$, $\mathcal{N}_A$, and $\mathcal{N}_S$ did not submit the report because they did not receive the $RREP$ packet. For

Case 5, the route was broken during relaying the first data packet, and node $\mathcal{N}_B$ is the last node that received the packet.

3.3.3   Cheaters Identification

As shown in Fig. 3.1, in the *Cheaters Identification* phase, Tp processes the cheating reports to identify the cheating nodes and correct the financial data. Our objective of securing the payment is to prevent the attackers (singular of collusive) from stealing credits or paying less, i.e., the attackers should not benefit from their misbehavior. We should also guarantee that each node can earn the correct payment even if the other nodes in the route collude to steal credits. In order to identify the cheaters, Tp requests the *Evidence* only from the node that claims more payment instead of all the nodes in the route because it should have the necessary and undeniable proofs (signatures and hash chain elements) for identifying the cheaters. *Evidence*s are not requested from the source and destination nodes when their reports are consistent because they can create them if they collude. By this way, *Tp can precisely identify the cheating nodes with requesting few Evidences and without false accusations.* To verify an *Evidence*, Tp computes the *Evidence*'s security token by generating the nodes' signatures and hashing them. The *Evidence* is credible if the resultant hash value is identical to the *Evidence*'s security token. Tp also verifies the source and destination nodes' hash chains by making sure that V_D^0 and V_S^1 are obtained from iteratively hashing V_D^X and V_S^X. The number of delivered packets (X) can be computed from the number of hashing operations required to obtain V_D^X from V_D^0, and the number of transmitted packets (X) can be computed from the number of hashing operations required to obtain V_S^X from V_S^1.

Without loss of generality, Table 3.4 gives numerical examples for cheating payment reports. The examples given in the table will clarify how cheating nodes can be precisely identified without false accusations. In Cases 1 and 2, the reports of the intermediate and destination nodes are consistent, but the source node claims sending fewer number of messages. The source node can compose valid *Evidence* if it cheats because it has the token V_D^6, and thus it is not effective to request the *Evidence* from the source node. Tp can request the *Evidence* from an intermediate node or the destination node. The source node is cheater if the *Evidence* is correct because it cannot be composed without the source node's hash value V_S^{10} and it is infeasible to compute this hash value. For Case 2, it is obvious that the route was broken at $\mathcal{N}_B$ during relaying the data packet number ten. For Case 3, the source and destination nodes' reports are consistent but the intermediate nodes claim relaying more messages. If an intermediate node submits valid *Evidence*, the source and destination nodes are cheaters because the *Evidence* should have V_S^{12} and V_D^{11} or V_D^{12}. It is not effective to request the *Evidence* from the source or the destination node because they can collude to compute valid *Evidence*. In Case 4, the reports of the intermediate and destination nodes are consistent but the source node claims sending more messages. This case may be rare because the rational attackers will attempt to steal credits or pay less. Tp can clear the payment according to the nodes' reports without requesting *Evidences* to achieve our security strategy and discourage submitting incorrect reports because the source node pays more if it lies and the other nodes lose credits if they lie. However, Tp can identify the cheating nodes by requesting the *Evidence* from the source node because it should contain V_D^{12} or V_D^{11}. If the *Evidence* is correct, the destination node is cheater but the intermediate nodes should not be punished because eight messages may be indeed relayed in the session and the source and destination nodes collude to falsely accuse the intermediate nodes. Case 5 is similar to Case 4 but the source and destination nodes report the same number of messages. The payment is cleared according to the nodes' reports to punish the nodes that submit incorrect reports without stealing credits.

In Cases 6 and 7, $\mathcal{N}_B$ can prove the credibility of its reports and earn the deserved payment even if the other nodes in the session collude. For Case 7, $\mathcal{N}_B$ claims delivering seven messages but the other nodes claim receiving seven messages and delivering only six messages. If $\mathcal{N}_B$ is honest, its *Evidence* should have V_D^7. If the *Evidences* of $\mathcal{N}_B$ are valid, the source and destination nodes are cheaters in Case 6 but only the destination node is a cheater in Case 7. For Case 8, as long as the destination node acknowledges receiving the message number seven, the intermediate nodes are rewarded for seven messages. In Cases 9 and 10,"––" means that the node did not submit the payment report. For Case 9, if $\mathcal{N}_A$ submits valid *Evidence*, the source and destination nodes are cheaters because they the route but did not submit the payment reports. $\mathcal{N}_B$ and $\mathcal{N}_C$ are not rewarded to discourage un-submitting the payment reports. For Case 10, the source and destination nodes are charged but the intermediate nodes are not rewarded without requesting *Evidences* in order to punish the nodes that do not submit reports.

3.3.4 Credit-Account Update

As shown in Fig. 3.1, *Credit-Account Update* phase receives the financial information to update the nodes' credit accounts according to the charging and rewarding policy discussed in Sect. 3.1. The financial information includes who pays whom and how much. The identities of the payers (source and destination nodes) and payees (intermediate nodes), the number of messages, and the ratio of the payment are included in the payment reports. For the cheating reports, the payment is cleared in such a way that prevents stealing credits and punishes the cheating nodes to discourage cheating actions. For example, Case 1 in Table 3.4 is cleared for ten messages if $\mathcal{N}_S$ cheats, and Case 3 is cleared for 12 messages if $\mathcal{N}_S$ and $\mathcal{N}_D$ cheat. For Cases 5, each node is rewarded or charged according to their reports, so that the payee that submits less payment is rewarded less, and the payers that submit more payment are charged more. For Case 9, $\mathcal{N}_A$ is rewarded for four messages, $\mathcal{N}_B$ and $\mathcal{N}_C$ are not rewarded because they do not submit the report, and $\mathcal{N}_S$ and $\mathcal{N}_D$ are charged for four messages. By this way, the nodes that submit incorrect payment reports always lose credits.

In receipt-based incentive schemes, a receipt can be cleared once it is submitted because it carries undeniable security proof, but in our scheme, Tp has to wait until receiving the reports of all nodes in a route to verify the payment. The payment clearance and trust update delay is the elapsed time from a route's establishment time until the payment is cleared and the trust values are updated. The maximum payment clearance delay (or the worst-case timing) occurs for the roues that are established shortly after at least one node contacts Tp and the node submits the report after the certificate lifetime (T_{Cert}). In this case at least one report is submitted after T_{Cert} from the route establishment. It is worth to note that the maximum time duration for a node's two consecutive contacts with Tp is T_{Cert} to renew its certificate to be able to use the network.

Figure 3.11 shows the worst-case timing of the submission and clearance of the reports with considering that the reports are submitted every T_{Cert}, where SUB_R, SUB_E, CLR_FR, and CLR_CR refer to the events of submitting reports, submitting *Evidences*, clearing fair reports, and clearing cheating reports. At t_1, the nodes submit the payment reports of the routes established in $[t_0, t_1)$ and the fair reports can be cleared. Thus, the maximum payment clearance delay of fair reports is T_{Cert} for the routes established shortly after t_0, but the average payment clearance delay is $T_{Cert}/2$ for the routes established in $[t_0, t_1)$ with assuming that the routes are established according to uniform random distribution. At t_2, Tp requests the *Evidences* of the cheating reports of the routes established in $[t_0, t_1)$. Thus, the maximum payment clearance delay for cheating reports is $2T_{Cert}$ for the routes established shortly after t_0, but the average payment clearance delay is $1.5T_{Cert}$ for the cheating reports of the routes established in $[t_0, t_1)$. The figure also shows that the maximum time for storing an *Evidence* is $2T_{Cert}$, e.g., for the reports of the routes established shortly after t_0. Moreover, at t_2, the nodes delete the *Evidences* of the routes established in $[t_0, t_1)$ because Tp must have cleared their reports. The payment clearance delay is bounded by T_{Cert} for fair reports and $2T_{Cert}$ for cheating reports because Tp has to wait one T_{Cert} for requesting the *Evidences*. It is obvious that each node has to store the *Evidences* for at most $2T_{Cert}$.

However, the nodes submit the reports at different times because the connection to Tp may not be available on a regular basis, and thus the time interval between each two submissions may not be the same and may be less than T_{Cert}. Hence, the maximum payment clearance delay may be less than T_{Cert}. Note that if a node does not submit a report in time period T_{Cert} from the route establishment, the payment is cleared such as Cases 9 and 10 in Table 3.4. In order to estimate the average and the maximum payment clearance delay, we assume that T_i is a continuous random variable that denotes the time duration between two submissions for a node, where $T_i \in [0, T_{Cert}]$ and the submission durations of the nodes are independent and identically distributed (i.i.d.) random variables. We consider two models for T_i. For model I, each node contacts Tp when it accumulates a large number of reports or when the remaining time of its certificate's lifetime is short to reduce

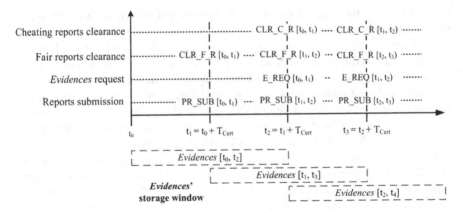

Fig. 3.11 The worst-case timing of the reports submission and clearance

the communication overhead. We model this behavior with truncated exponential distribution given in Eq. (3.1), where the probability that a node contacts Tp is high as T_i approaches T_{Cert}. For model II, a node submits the reports once it has a connection to Tp and the connections are uniformly distributed over the time interval $[0, T_{Cert}]$.

Equations (3.2) and (3.3) give the probabilities of submitting the reports at most by time t for models I and II, respectively. The payment of a route is cleared when all the nodes in the route submit their reports. $T(n)$ is a continuous random variable that denotes the time duration for n nodes to contact Tp, where $T(n) \in [0, T_{Cert}]$. Equation (3.4) gives the probability that $T(n)$ is at most t. The probability density functions of $T(n)$ using models I and II are given in Eqs. (3.5) and (3.6), respectively. Equation (3.7) gives the average time duration for n nodes to contact Tp, which is equivalent to the maximum payment clearance delay for a route with n nodes. Equation (3.8) gives the average payment clearance delay for a route with n nodes with assuming that the routes are established according to uniform distribution.

$$f(T_i) = \frac{\lambda \exp^{-\lambda t}}{1 - \exp^{-\lambda T_{Cert}}} \tag{3.1}$$

$$P(T_i \le t) = \frac{1 - \exp^{-\lambda t}}{1 - \exp^{-\lambda T_{Cert}}} \tag{3.2}$$

$$P(T_i \le t) = \frac{t}{T_{Cert}} \tag{3.3}$$

$$P(T(n) \le t) = \prod_{i=1}^{i=n} P(T_i \le t) \tag{3.4}$$

$$f(T(n)) = \exp^{-\lambda t} \lambda n \frac{(1 - \exp^{-\lambda t})^{n-1}}{(1 - \exp^{-\lambda T_{Cert}})^n} \tag{3.5}$$

$$f(T(n)) = \frac{n}{T_{Cert}} (\frac{t}{T_{Cert}})^{n-1} \tag{3.6}$$

$$E(T(n)) = \int_0^{T_{Cert}} t f(T(n)) \; dt \tag{3.7}$$

$$P_C(n) = \frac{E(T(n))}{2} \tag{3.8}$$

$$E(T_i) = \frac{1}{\lambda} [\frac{1 - (\lambda T_{Cert} + 1) \exp^{-\lambda T_{Cert}}}{1 - \exp^{-\lambda T_{Cert}}}] \tag{3.9}$$

We have selected λ to be $\frac{1}{7}$ and T_{Cert} to be 10, 15, and 20 days, to make the average time duration between two report submissions (given in Eq. (3.9)) 3.9, 4.9, and 5.73 days, respectively. Figures 3.12 and 3.13 show the probability that the payment of a route with n nodes is cleared at most by time t for models I and II, respectively, with T_{Cert} of 15 days and λ of $\frac{1}{7}$. The figures show that the increase of n decreases the probability of clearing the payment by time t because Tp has to wait more time to receive reports from more nodes. Moreover, the maximum payment clearance delay is less than T_{Cert} with high probability. Figures 3.14 and 3.15 show the average payment clearance delay at different values of n for models I and II, respectively. It can be seen that the average payment clearance delay can be much less than T_{Cert}. For example, at $n = 5$ nodes and $T_{Cert} = 15$ days, the average

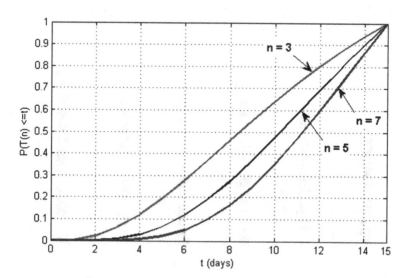

Fig. 3.12 $P(T(n) \leq t)$ versus t at different values of n for model I

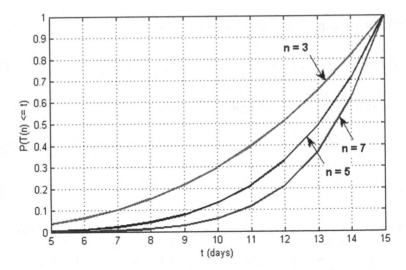

Fig. 3.13 $P(T(n) \leq t)$ versus t at different values of n for model II

payment clearance delay is 5 and 6.2 for models I and II, respectively. It can also be seen that shortening T_{Cert} can decrease the payment clearance delay.

Although our analysis demonstrates that the payment clearance delay can be less than T_{Cert} when the nodes submit their reports according to truncated exponential and uniform distributions, this delay is bounded by T_{Cert}. The delay is acceptable due to using post-paid payment model where the nodes communicate first and pay later, i.e., the nodes do not need to wait until clearing the payment to communicate. Moreover, the nodes can purchase credits for real money.

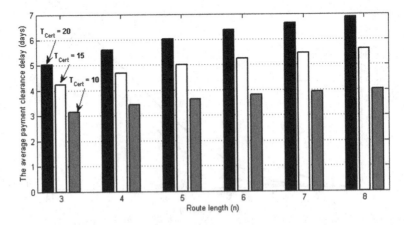

Fig. 3.14 The average payment clearance delay for model I

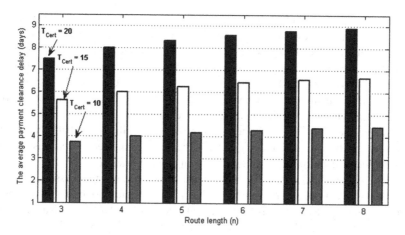

Fig. 3.15 The average payment clearance delay for model II

3.4 Security Analysis

Our security objective is to prevent the attackers from achieving gains such as stealing credits or paying less. In our incentive scheme, the charges and rewards are determined based on payment reports submitted by autonomous nodes, so a node or even a group of colluding nodes may attempt to cheat to increase their welfare.

To simplify our presentation, we considered that a keyed hash value covers only the message, but for better security, it should cover the whole packet. For example, in Fig. 3.7, the keyed hash value of $\mathcal{N}_B$ should be $H_{K_{SB}}(M_X, V_S^{N-X+1}, H_{K_{SC}}(M_X), H_{K_{SD}}(M_X))$, so if $\mathcal{N}_A$ manipulates the hash value of $\mathcal{N}_D$, e.g., to consume the nodes' resources because the packet will be dropped at $\mathcal{N}_D$, $\mathcal{N}_B$ can stop propagating the incorrect packet.

Since the source node attaches a keyed hash value for each node in the route, it is obvious that the packet overhead will be large for long routes. To reduce the packet overhead, the message's keyed hash value can be truncated significantly, e.g., the size of the truncated hash value (η) can be 4 or 5 bytes instead of 16 bytes in *HMAC-MD5*. This severe hash truncation is secure in ESIP for the following reasons:

1. The packet security lifetime is extremely short, i.e., if an intermediate node does not relay a packet in a short time, the route is considered broken and re-established, so a malicious node does not have long time to run complicated algorithms to figure out the truncated keyed hash values of the manipulated message.
2. Without knowing the secret key, computing the keyed hash value is infeasible.
3. The attacker has to figure out a keyed hash value for each victim between its device and the other colluder.

Therefore, the attacker has to compute multiple truncated keyed hash values without knowing the keys in a limited time, which is infeasible. What an attacker can do is to replace a truncated keyed hash value with a random value, but the probability to hit the correct value is extremely low, e.g., for $\eta = 4$ bytes, the probabilities to hit one and two correct hash values are 0.23×10^{-9} and 0.05×10^{-18}, respectively. Moreover, if the manipulated hash value is not correct, the attacker's neighbor drops the packet, and thus the attackers' trust values degrade.

However, the hash truncation increases the random collision probability, i.e., the corrupted and the original messages have the same truncated keyed hash value. Using birthday paradox, the random collision probabilities for η of 4 and 5 bytes are 1.2×10^{-5} and 7.63×10^{-7}, respectively. In addition, since the message integrity is checked at each node, the probability that the destination node falsely accepts a corrupted message as correct is $(1.2 \times 10^{-5})^{n_1}$ for η of 4. This is equivalent to the probability that the hash collision occurs at n_1 successive nodes, where n_1 is the number of nodes from the node at which the message is corrupted to the destination node. This probability is very small and can be reduced with the increase of η, but the packet overhead increases. Therefore, the size of η can be variable to balance the probability of falsely accepting corrupted message and the packet overhead, i.e., η can be longer for short routes. Moreover, some nodes in a route can have longer η than others, e.g., η can be longer for the destination node to prevent falsely accepting corrupted messages. *MD5* is faster and has shorter digest length than *SHA-1* but *SHA-1* is more collision resistant, so *SHA-1* can be used in digital signing operations that require high collision resistance and *MD5* is used to compute the keyed hashes and the hash chains.

For *Free-Calling* (or *Riding*) attacks, two colluding intermediate nodes in a route may manipulate the packets by adding their data to communicate freely. To thwart this attack, the messages' integrity and authenticity are checked at each node by verifying the message's keyed hash value, and thus the first intermediate node after the attacker can detect any addition or modification to the packets and drop them. Launching this attacks against the *ACK* packets is infeasible because each *ACK* has a hash-chain element that is verified at each node, and thus the packet's integrity and authenticity can be verified at each node. The integrity and authenticity of the route discovery packets can also be verified at each node, as will be discussed in Sect. 4.3.

For *One-Redemption* attack, the attackers attempt to make the payment reports of different routes identical to pay once. This is because Tp clears the payment of a route only once. In RACE, the reports cannot be repeated because each report has a unique identifier that consists of the identities of the nodes in the route and time stamp. This identifier changes once the route or the time changes. The reports are different even if the same nodes participate in different routes at different times. Even if the attackers could establish different routes at the same time, the reports' identifiers are different because at least one intermediate node is different. For *Multiple-Redemptions* attack, the attacker attempts to deceive Tp by submitting the same payment reports multiple times to be rewarded several times for relaying the same packets. Tp can thwart the attack and identify the attacker because before processing a report, it uses its identifier to ensure that the report has not been processed before.

For *Evidence-Forgery/Manipulation* attack, the attackers attempt to forge an *Evidence* for communication sessions that did not happen or manipulate a valid *Evidence* to steal credits. This is infeasible with using secure hash function and secure public-key cryptography because it is infeasible to compute V_D^X from V_D^{X-1} or V_S^X from V_S^{X-1}. It is also infeasible to modify or forge the source/destination nodes' signatures, compute the private keys from the public ones, and compute the hash value of the signatures without computing the signatures. Moreover, if an attacker attaches a random value for an *Evidence*'s security token, the probability to hit the correct value is nearly zero, e.g., this probability is 6.84×10^{-49} with using *SHA-1* [27, 28] with digest value of 20 bytes. Tp can identify the attackers when their *Evidences*' verifications fail. The intermediate nodes can verify the source and destination nodes' signatures and hash chain, which is important to make sure that they can compose valid *Evidences* to secure the payment. Without these verifications, the source node may send packets with invalid payment tokens to avoid paying.

For *Impersonation* attack, the attackers attempt to impersonate others to communicate freely or steal credits. This attack is infeasible because the nodes use their private keys in signing the packets and use their secret symmetric keys in submitting the payment reports, and the attackers cannot compute others' private keys. For *Packet-Replay* attack, the attackers record valid packets and replay them in different place and/or time claiming that they are fresh to establish routes under the name of others to communicate freely. In our scheme, a fresh time stamp is used in the *RREQ* packet to establish a route and thus stale *RREQ* packets can be identified and dropped when the time stamp is not within a proper range.

The attackers may attempt to manipulate their reports to launch *Reduced-Payment* and *Fake-Intermediate-Node* attacks. For *Reduced-Payment* attack, some intermediate nodes may collude with the source and destination nodes to submit payment reports with less payment. For example, if ζ intermediate nodes launch this attack successfully in a route with n nodes, the colluders can save $(n - 2 - \zeta)(X - \omega)\lambda$ credits, where X and ω are the correct and the submitted number of messages, respectively. The source and destination nodes can compensate the colluding intermediate nodes. In our scheme, even if a group of nodes colludes to reduce the rewards of an honest node, the node can compose valid *Evidence* and earn the correct payment, such as Cases 6 and 9 in Table 3.4.

For *Fake-Intermediate-Node* attack, the attackers may claim that they have non-existent neighbors to let the source and destination nodes pay more, collect credits to the non-existent neighbor, or falsely accuse the non-existent node of breaking the route or un-submitting the payment report. As will be discussed in Sect. 4.3, the nodes authenticate themselves in route discovery phase and the authentication code is attached to the *Evidence*. Moreover, in our scheme, if the victim nodes are intermediate, the payment is cleared without punishing or rewarding them as discussed in Cases 9 and 10 in Table 3.4, but if the victim nodes are source or destination, the attackers are evicted because they cannot submit correct *Evidences*.

Table 3.4 Numerical examples for cheating payment reports

Case number		$\mathcal{N}_S$	$\mathcal{N}_A$	$\mathcal{N}_B$	$\mathcal{N}_C$	$\mathcal{N}_D$
1	X	6	10	10	10	10
	F	1/0	1	1	1	1
2	X	6	10	10	9	9
	F	1/0	0	0	1	1
3	X	5	12	12	12	5
	F	1	1/0	1/0	1/0	1
4	X	12	8	8	8	8
	F	1/0	1	1	1	1
5	X	9	4	4	4	9
	F	1/0	1	1	1	1/0
6	X	14	14	22	14	14
	F	1	1	1/0	1	1
7	X	7	7	7	7	6
	F	0	0	1	0	1
8	X	7	7	7	7	7
	F	0	0	1	0	1
9	X	–	4	–	–	–
	F	–	1/0	–	–	–
10	X	6	–	–	–	6
	F	1/0	–	–	–	1/0

However, colluding nodes may exchange their cryptographic information to insert non-existent nodes to collect credits for each other without relaying packets. This attack is a type of the known routing attack called *Route-Lengthening*. First, the widespread of exchanging the cryptographic information is unlikely because colluders can steal the credits of each other or commit malicious actions under their names. Second, the attack does not always work because it may lead to sub-optimal route due to the preference of the shortest routes. Third, Tp can identify the attackers when it observes that some nodes appear in different locations at the same time. Finally, the proposed solutions for secure routing protocols such as ARAN [29] and *Ariadne* [30] can be implemented with our scheme.

For *Destination-Node-Robbery* attack, the source node colludes with some intermediate nodes to steal credits from the destination node by sending bogus messages paid by the source and destination nodes. For example, if the source node colludes with κ intermediate nodes, the colluding intermediate nodes earn $X \lambda \kappa$ credits for relaying X packets, but the source node pays $X \lambda n P_r$. Obviously, the colluders can achieve profits when $(X \lambda \kappa - X \lambda n P_r) > 0$ or $(\kappa - n P_r) > 0$. In our scheme, the attackers do not earn credits from sending bogus messages because the intermediate nodes are rewarded only when the destination node acknowledges receiving correct messages. A route cannot be established and a valid *Evidence* cannot be composed if the destination node is not interested in the communication because its signature is required.

For *Message-Repudiation* attack, the attackers attempt to deny transmitting a message. In our scheme, each node can ensure that the intended user has sent a message, but unlike signature-based protocols, it cannot prove that to a third party. However, message non-repudiation is important for other applications such as electronic commerce where a user sends messages to authorize the recipient to perform actions on its behalf. For *Payment-Repudiation* attacks, the attackers attempt to deny establishing a route or the amount of payment so as not to pay. In our scheme, the payers cannot deny the payment because the signatures and the hash chains can guarantee the payment non-repudiation.

Without proper payment model, the rational attackers may try to cheat to increase their welfare. Our charging and rewarding policy has been developed to counteract the rational cheating actions and encourage the nodes' cooperation. Particularly, a rational node can exhibit one of the following actions:

1. If the intermediate nodes are rewarded for relaying the messages that do not reach the destination node such as the schemes in [8, 10], colluding intermediate nodes can increase their rewards with consuming low resources by dropping a message and relaying only the security tag (hash chains' elements) that is much shorter than the message to claim the payment for relaying the message. This attack can be successful because the security tag is enough to compose a valid *Evidence*. To prevent this cheating action, our payment model encourages the nodes to rely the messages because they are rewarded only when the destination node acknowledges receiving them. Moreover, the attackers' trust values degrade when they drop packets and thus their chance to be involved in routes decreases.
2. If the source and destination nodes are charged only for the successfully delivered messages, the destination node may receive a message but does not send *ACK* so as not to pay, or a colluding intermediate node claim that the packet is not delivered because the route is broken. To prevent this cheating action, both the source and destination nodes are charged for the un-delivered messages.

Relaying the route establishment packets is beneficial for the nodes to participate in routes and thus earn credits. If the destination node does not acknowledge a message, it pays for the message and the source node does not send the next message. Relaying the *ACK* packets is beneficial for the intermediate nodes to trigger the source node to generate more messages and thus earn more credits.

Similar to most of the existing schemes, we believe that *ACK* packets not only can achieve reliable packet transmission but also are essential to secure the payment whether only the source node pays or both the source and destination nodes pay. This is because: (1) if only the source node pays, some intermediate nodes may drop the messages and save only the source node's signature to be rewarded for the undelivered messages; and (2) if both the source and destination nodes pay, the source node may collude with some intermediate nodes to steal credits from the destination node by sending bogus messages.

Although the charges are always more than or equal to the rewards, our payment model does not make the credits disappear because purchasing credits with real money can compensate the credit loss. The rich nodes that have much more credits than their credit consumption may cease relaying others' packets to save their resources. *Tp* converts credits to real money to motivate these nodes to cooperate.

In addition, trust values will be used to encourage the rich node to cooperate, as will be discussed in Chap. 4.

In our payment model, the source and destination nodes can communicate even if they do not have sufficient credits at the communication time. Contacting Tp in each session to verify the nodes' accounts will incur too much overhead. For *Payment-Denial* attacks, the attackers may join the network for short time and leave without paying their debts. Different from the traditional ad hoc networks that can be temporarily deployed and similar to the current cellular networks, *MWN*'s lifetime is long and the nodes have long-term relations with the network. The post-paid payment policy has been widely used in many services successfully such as credit cards and cellular networks. In our scheme, each node needs a certificate to participate in the network, and issuing a certificate is not free to make changing identity costly. Moreover, similar to the existing cellular networks, Tp stores the personal and financial information of the users and can take the legal procedures against the users who do not pay. To limit overspending, a node's certificate lifetime can be short and depend on the node's available credits at the certificate issuing time and its average credit consumption rate. The pre-paid and post-paid payment models can also be jointly used to reduce the debt, e.g., each node should have a minimum amount of credits when renewing the certificate.

3.5 Performance Evaluations

In this section, simulation results are given to evaluate the overhead and the expected performance of the proposed protocols and schemes. We first evaluate ESIP communication protocol and then we evaluate the reports' submission and clearance overhead in RACE.

3.5.1 Replacing Signatures with Hashing Operations

3.5.1.1 Simulation Setup

For the public key cryptography, we use 1,024-bit RSA and 128-bit DSA with signature tags of 128 and 40 bytes, respectively. According to NIST guidelines [31], the secure private keys should have at least 1,024 bits in RSA and 128 bits in DSA. For the hash functions, we use MD5 and HMAC-MD5 [27] with digest length of 16 bytes and SHA-1 hash function with digest length of 20 bytes [27, 28]. For the pairing operation, we consider the Tata pairing implementation on MNT curves where $\mathcal{G}$ is represented by 171 bits, and the order $\mathcal{P}$ is represented by 170 bits. The discrete logarithm in $\mathcal{G}$ is as hard as the discrete logarithm in $\mathcal{Z}_P^*$ where $\mathcal{P} = 1,024$ bits. The network simulator NS2 is used to implement ESIP and signature-based incentive schemes.

We simulate multi-hop wireless network by randomly deploying 35 mobile nodes in a square cell of 1000×1000 m. The nodes' radio transmission range is 125 m and the data transmission rate is 2 Mbits/s. The Distributed Coordination Function (DCF) of IEEE 802.11 is implemented as the medium access control (MAC) layer protocol. To emulate the node mobility, we adopt the modified random waypoint model [32]. Specifically, a node travels towards a random destination uniformly selected within the network field; upon reaching the destination, it pauses for some time; and the process repeats itself afterwards. The node speed and the pause time are uniformly distributed in the ranges [0, 10] m/s and [0, 50] s, respectively. The constant bit rate (CBR) traffic source is implemented in each node, and the source and destination pairs are randomly chosen. All the data packets are 512 bytes and sent at the rate of 2 packets/s. The time stamp and an identity are five and four bytes, respectively. Each simulation is performed 50 runs, and each run is executed for 15 simulated minutes. The averaged results are presented with 95 % confidence interval.

In order to estimate the expected computational times of the cryptosystems required in our schemes, we have implemented them using $Crypto++5$ library [33] and a laptop with an Intel processor at 1.6 GHz and 1 GB RAM, and the measurements are given in Table 3.5. The measurements indicate that the signing and verifying operations require 15.63 ms and 0.53 ms, respectively for RSA. Moreover, the signing and verifying operations require 7.94 ms and 9.09 ms, respectively for DSA. A concern in using DSA for multi-hop networks is that the verifying operations performed by the intermediate and destination nodes require more times than the signing operations performed by the source node, and a concern in using RSA is its longer signature tag. The measurements also indicate that the computational times for hashing a 512-byte message and performing a pairing operation are 8.56 µs and 4.34 ms, respectively. It can see that the computational times of the signing and verifying operations are much less than this of the hashing operations. For example, the computational times of the signing and verifying operations are sufficient for 539 and 18 512-byte hashing operations, respectively with using RSA.

Moreover, the energy consumption of the used cryptosystems are measured in [34, 35] and the results are given in Table 3.5. It can be seen that the consumed energy for signing and verifying operations with using RSA are sufficient for 1,404 and 41 512-byte hashing operations, respectively. The resources of a real mobile node may be less than a laptop, so the results given in Table 3.5 are scaled with the factor of the five in our simulations to estimate a limited-resource node.

In Table 3.6, the statistics of the route length and the network connectivity of the simulated network are given. $P(R_L \leq 4)$ is the probability that a route has at most four nodes including the source and destination nodes. The network connectivity is the ratio of the connected routes to the total number of possible routes with assuming that any two nodes can be the source and destination pair. The statistics indicate that the simulated network is well connected and the route length is acceptable.

Table 3.5 The processing times and energy of the used cryptosystems

		Processing time	Processing energy
1024-bit RSA	Signing operation	15.63 ms	546.5 mJ
	Verifying operation	0.53 ms	15.97 mJ
128-bit DSA	Signing operation	7.94 ms	313.6 mJ
	Verifying operation	9.09 ms	338.02 mJ
Pairing operation		4.34 ms	25.5 mJ
MD5		8.56 μs/512 bytes	0.302 mJ
SHA-1		29 μs/512 bytes	0.76 μJ/B

Table 3.6 Statistics of the simulated network

Average network connectivity	$P(R_L \leq 4)$	$P(4 < R_L \leq 6)$	$P(6 < R_L \leq 8)$	$P(8 < R_L < 10)$	$P(R_L > 10)$
0.888	0.559	0.297	0.118	0.023	0.0041

3.5.1.2   Simulation Results

A. Average Packet Overhead

The average security packet overhead is defined as the average security related data relayed in all the hops of a route. In Fig. 3.16a, the security packet-overhead of signature-based incentive schemes is due to fixed-size and route-length-independent signature, e.g., 40 and 128 bytes for DSA and RSA based schemes, respectively. However, in Fig. 3.16b, the security packet overhead in ESIP is due to the 16-byte hash chain's element (V_S^X) and the message hash series $HS(M_X)$ with η-byte truncated hash values when $X > 1$. The figure also shows that the security packet-overhead is reduced by η bytes at each hop because each node drops its message hash value. Unlike signature-based incentive schemes, the security packet-overhead of ESIP depends on the route length (R_L).

Figure 3.17 gives the relation between the average security packet-overhead and the route length in ESIP. The figure shows that *even at unrealistic and extreme cases, e.g., $R_L = 20$ nodes, the average security packet-overhead is less than 55 bytes at $\eta = 4$ bytes*. Figure 3.18 gives the equivalent route lengths of signature-based incentive schemes and ESIP for the same average security packet overhead at $\eta = 4$ bytes. For example, the routes with six nodes in DSA and RSA based schemes are equivalent to routes with 8 and 15 nodes in ESIP for the same average security packet-overhead, respectively.

The figure shows that the average security packet overhead in ESIP is less than that of the DSA and RSA based schemes when $R_L < 13$ nodes and $R_L < 24$ nodes, respectively at $\eta = 4$. Moreover, the security packet overhead of ESIP is less than that of the DSA-based incentive schemes when R_L is fewer than 17 and 10 nodes for η of 3 and 5 bytes, respectively. The security packet overhead of ESIP is also less than that of the RSA-based incentive schemes when R_L is fewer than 75 and 45 nodes for η of 3 and 5 bytes, respectively. Although DSA has less signature size than

RSA, it much increases the end-to-end packet delay due to its longer verification time, as will be discussed in Sect. 3.5.1.2B.

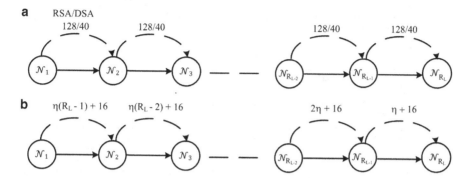

Fig. 3.16 The hop-by-hop security packet-overhead of ESIP and signature-based incentive schemes. (**a**) The hop-by-hop security packet-overhead in RSA/DSA based incentive schemes. (**b**) The hop-by-hop security packet-overhead of ESIP

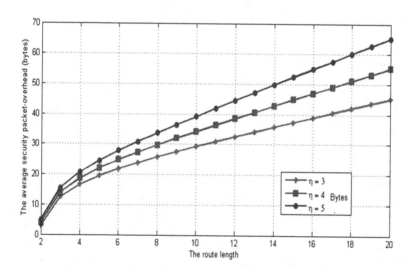

Fig. 3.17 The average packet security overhead in ESIP

Figure 3.19 shows the distribution of route length at different number of nodes in the simulated network. At 15 nodes, the network is lightly connected because the average connectivity is 0.66. As shown in Fig. 3.19a, 86 % of the routes have four nodes or fewer, and only 0.0238 % of the routes are longer than ten nodes.

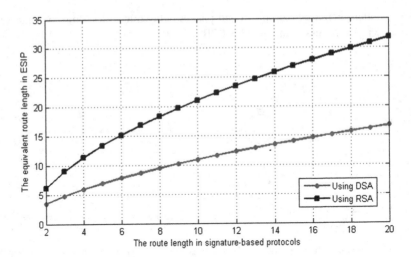

Fig. 3.18 The equivalent route lengths for the same security packet overhead

Table 3.7 The average connectivity, route length at different network parameters

Network dimension	Nodes' number	Average connectivity	Average route length	$Pr(R_L > 13)$
800 × 800	15	0.66	3.25	0
	30	0.97	3.66	0
	60	1	3.41	0
1600 × 1600	40	0.2235	3.6892	0.000444
	60	0.5394	5.5683	0.011
	100	0.9531	6.3174	0.0059
2000 × 2000	100	0.5591	7.4081	0.091
	150	0.948	7.7624	0.0539
	200	0.992	7.172	0.01225

At 35 nodes, the average network connectivity is 0.99, the probability that a route is shorter than seven is 99.7 %, and the probability that a route is longer than ten nodes is 0.0151 %. At 50 nodes, the probability that a route is shorter than seven is 98.1 %, and the probability that a route is longer than ten nodes is negligible. For dense network with 100 nodes, the probability that a route is shorter than seven is 99.99 %, and the probability that a route is longer than ten nodes is negligible. Table 3.7 gives the probability that a route is longer than 13 nodes ($P(R_L > 13)$) at different network dimensions. The conclusion of these results is that *the route length is less than 13 nodes with very high probability under realistic network parameters, and thus the expected security packet overhead of ESIP is less than those of the DSA and RSA based incentive schemes.*

The average packet overhead is the average additional data (in bytes) attached to the message including the routing and security data. Table 3.8 gives the average packet overhead in ESIP and signature-based incentive schemes. The packet overhead with using RSA is much longer than that with using DSA due to its longer signature. For signature-based incentive schemes, the average packet overhead in the first packet is longer than those of the next packets due to attaching the source node's certificate. For the first packet, the average packet overhead of ESIP is more than that of the signature-based incentive schemes due to attaching the source node's signature, V_S^N and $HS(M_1)$. However, the packet overhead is less in the next packets because the source node does not attach signatures. For the first packet, the packet overhead of ESIP is 1.18 and 1.067 times the overhead of DSA and RSA based incentive schemes, and for a series of two packets, the ratios become 0.98 and 0.79, so *from the second packet, we start to gain the revenue from the overhead investment in the first packet.* Moreover, *for a series of ten packets, the data packet overhead in ESIP is 70 % and 37 % of those in the DSA and RSA based incentive schemes,* respectively.

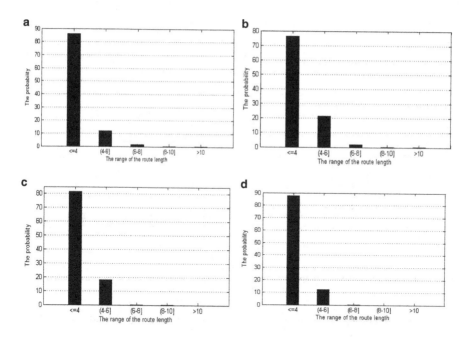

Fig. 3.19 Route length distribution. (**a**) At 15 nodes. (**b**) At 35 nodes. (**c**) At 50 nodes. (**d**) At 100 nodes

B. Average End-to-End Packet Delay

The required cryptographic operations for ESIP and signature-based incentive schemes are given in Table 3.9, where P, V, S, and H refer to pairing, verifying, signing, and hashing operations, respectively. It can be seen that ESIP requires more

Table 3.8 The average data packet overhead (bytes)

		RSA	DSA
Signature-based	First packet	279	103
incentive schemes	Subsequent packets	143	55
ESIP	First packet	297.73	121.73
	Subsequent packets	33.73	33.73

Table 3.9 The required cryptographic operations in ESIP and signature-based incentive schemes

	ESIP			Signature-based incentive schemes		
	Source	Intermediate	Destination	Source	Intermediate	Destination
First data packet	$S + HR_L$	$2V(R_L - 2) +$ $2H(R_L - 2)$	$2V + H$	S	$2V(R_L - 2)$	$2V$
Subsequent data packets	HR_L	$2H(R_L - 2)$	H	S	$V(R_L - 2)$	V

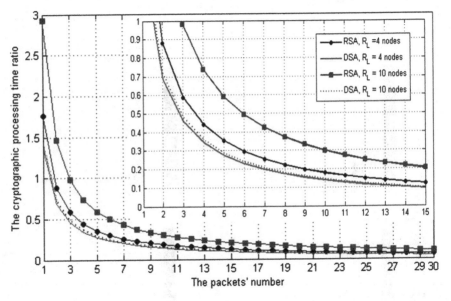

Fig. 3.20 The ratio of ESIP's cryptographic delay to that of signature-based incentive schemes

cryptographic operations in the first data packets, but from the second data packet, ESIP requires only hashing operations.

The cryptographic delay ratio is the required cryptographic computation time in ESIP to that of the signature-based incentive scheme. For N messages, $2N$ signatures are generated from the source and destination nodes in signature-based incentive schemes, but only two signatures are generated in ESIP. The cryptographic delay ratio versus the number of messages is shown in Fig. 3.20. For the first packet and R_L of 4 nodes, it can be seen that the cryptographic delay ratio is 1.4 and 1.75 with using DSA and RSA, respectively, and the ratio drops to 0.68 and 0.88 for

Table 3.10 The average packet series size, and cryptographic time and energy ratios

Node speed			[0, 2] m/s		[0, 10] m/s	
Nodes' number			15	35	15	35
0.5 packet/s	Average packet series size		126.8	134.15	42.55	40.425
		DSA	0.029	0.015	0.09	0.098
	Cryptographic energy ratio	RSA	0.033	0.018	0.1	0.11
		DSA	0.037	0.019	0.117	0.13
	Cryptographic delay ratio	RSA	0.051	0.027	0.16	0.173
1 packet/s	Average packet series size		289.7	258	94.6	95.4
		DSA	0.01	0.012	0.05	0.05
	Cryptographic energy ratio	RSA	0.011	0.015	0.055	0.056
		DSA	0.011	0.015	0.063	0.064
	Cryptographic delay ratio	RSA	0.016	0.022	0.084	0.088

two packets. Moreover, *for 13 packets, ESIP requires only 10 % and 12 % of the cryptographic delay in DSA and RSA based incentive schemes, respectively.* The reason for this big reduction is that the computation time of the hashing operation is negligible comparing to these of the signing and verifying operations. Therefore, the cryptographic delay ratio is nearly $\frac{2}{2N}$, i.e., the reciprocal of the packets' number. In addition, the simulation results given in Table 3.10 demonstrate that under different network parameters, the average size of the packet series is greater than 13, and the cryptographic delay in ESIP is much less than those of DSA and RSA based incentive schemes.

The average end-to-end packet delay is the average time for the packets to traverse the network from the source node to the destination node. Figure 3.21 shows the average end-to-end packet delay in ESIP and the signature-based incentive schemes at different traffic load expressed in number of connections, and Table 3.11 gives the confidence intervals of Fig. 3.21b. The simulation results demonstrate that *ESIP can significantly reduce the average end-to-end packet delay comparing to the DSA and RSA based incentive schemes because the hashing operations that are computationally free (43 µs per operation) dominate the nodes' operations.* Up to 20 connections, the cryptographic delay dominates the channel contention and queuing delays, but over 20 connections, the delay significantly increases with and without the incentive scheme because the channel contention and queuing delays dominate. Although DSA has shorter signature than RSA, it causes longer delay in the signature-based incentive schemes due to its longer verification time. However, DSA increases the delay slightly in ESIP because the effect of the long delay of the first packet vanishes with the dominant hashing operations. Hence, *ESIP can be implemented more efficiently using DSA because it has shorter signature and the hashing operations can mitigate the long delay of the first packet.*

C. Packet Delivery Ratio

The packet delivery ratio is the average ratio of the number of messages that successfully delivered to the destination nodes to those sent by the source nodes. Figure 3.22 gives the packet delivery ratio for ESIP and the original DSR at

Table 3.11 Ninety-five percent confidence interval (C.I.) for mean

Connections' number	C. I. for mean	End-to-end delay			Packet delivery ratio		
		ESIP with DSA	ESIP with RSA	DSR	ESIP with DSA	ESIP with RSA	DSR
	Upper limit	23.1	22.8	22.05	99.9	99.99	99.997
12	Mean	23.03	22.6	22	99.93	99.95	99.994
	Lower limit	22.96	22.4	21.95	99.9	99.93	99.991
	Upper limit	28.61	27.68	26.2	99.92	99.96	99.988
16	Mean	28.01	27.36	26	99.9	99.95	99.982
	Lower limit	27.41	27.04	25.8	99.88	99.94	99.982
	Upper limit	32.39	31.82	29.86	98.56	98.93	99.8
20	Mean	32.32	31.6	30	98.5	98.9	99.6
	Lower limit	32.25	31.38	30.14	98.44	98.87	99.4

different number of connections and Table 3.11 gives the confidence intervals. Up to 20 connections, the packet delivery ratio is quite high (above 98 %). Above 20 connections, the packet delivery ratio starts to decrease because more packets are dropped due to the increase in the number of congested nodes and packet collision. Since each node has only 50-packet queue size and increasing the connections' number increases the packet arrival rate, the node is congested and drops packets once the buffer is full. Moreover, increasing the cryptographic delay causes more congested nodes due to increasing the packet processing time. Comparing to the original DSR protocol, ESIP has a very little effect on the packet delivery ratio because the dominant hashing operations require very little computation time.

D. Average Network Throughput

The average network throughput is the amount of data received by all the nodes over the simulation time. Since the end-to-end packet delay and the packet delivery ratio in ESIP are close to those of the DSR, it is expected that the throughput is close as well. The simulation results shown in Fig. 3.23 demonstrate that ESIP has very little effect on the throughput comparing to the original DSR protocol. Up to 20 connections, the throughput increases with the increase of the number of connections, but above 20 connections, the increasing rate starts to decrease because the network reaches its capacity. As discussed in Sects. 3.5.1.2B and 3.5.1.2C, above 20 connections, the packet delivery ratio decreases and the end-to-end packet delay increases.

E. Energy Consumption

Energy is consumed in relaying packets and executing the cryptographic operations. As discussed in Sect. 3.5.1.2A, ESIP can reduce the packet overhead with a very high probability. From Table 3.5, it can be seen that the consumed energy for hashing operations is much less than those of the signing and verifying operations, which supports our approach of replacing signatures with hashing operations. Figure 3.24 shows the ratio of the required cryptographic energy in ESIP to those of the DSA and RSA based incentive schemes versus the number of data packets.

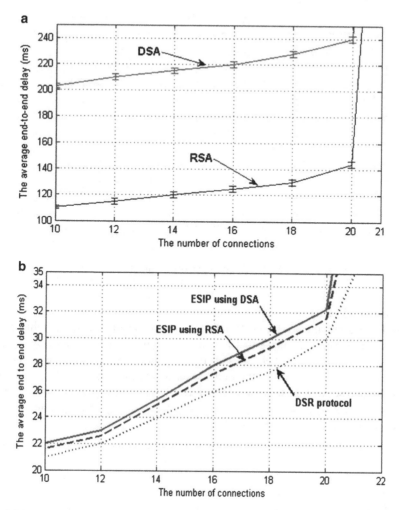

Fig. 3.21 The average end-to-end packet delay. (**a**) Signature-based incentive schemes. (**b**) ESIP and original DSR

For the first packet and R_L of 4, ESIP requires 1.025 and 1.175 of the consumed cryptographic energy in the DSA and RSA based incentive schemes, respectively. However, from the second packet ESIP requires less cryptographic energy, e.g., *for 10 packets, ESIP requires around 10 % of the cryptographic energy consumed in the DSA and RSA based incentive schemes at R_L of 4*. In addition, the simulation results given in Table 3.10 demonstrate that the average cryptographic energy consumed in ESIP is much less than those consumed in the DSA and RSA based incentive schemes.

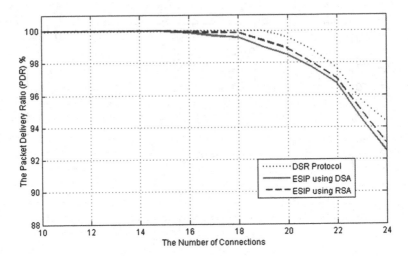

Fig. 3.22 The packet delivery ratio versus the number of connections

F. Effect of Mobility

At high node mobility, Table 3.10 indicates that the average cryptographic delay increases and Fig. 3.25 shows that the end-to-end packet delay increases. This is because the size of the packet series decreases at high node mobility, and thus the effect of the high overhead of the first packet increases. However, the simulation results demonstrate that the overhead of ESIP is still much less than those of the DSA and RSA based incentive schemes because only the efficient hashing operations are used after the first packet.

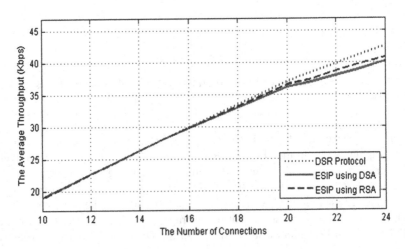

Fig. 3.23 The average throughput versus the number of connections

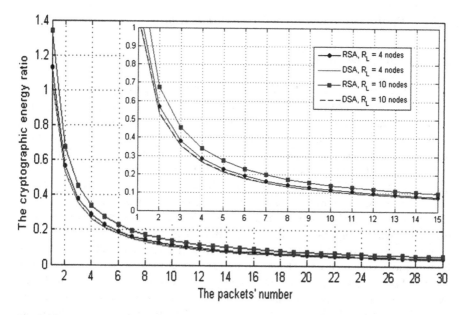

Fig. 3.24 The ratio of ESIP's cryptographic energy to that of the signature-based incentive schemes

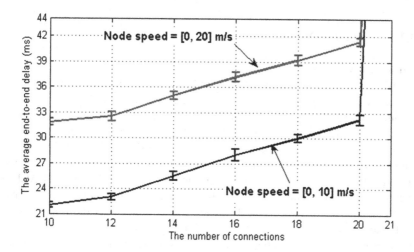

Fig. 3.25 The impact of node mobility on the end-to-end packet delay

3.5.2 Replacing Receipts with Payment Reports

3.5.2.1 Simulation Setup

We simulate a multi-hop wireless network by randomly deploying 50 mobile nodes in a square cell of 1200×1200 m. The nodes' transmission range is 150 m. The constant-bit-rate traffic source is implemented in each node and the source and destination nodes are randomly chosen. We use the modified random waypoint model [30] to emulate node mobility. The nodes' speed is uniformly distributed in the ranges $[0, 5]$ m/s and $[0, 10]$ m/s and the pause time is 20 s. The data transmission rate is 0.5 packet per second. The time stamp (Ts), node's identity (ID_i), and message number (X) are five, four, and two bytes, respectively. The size of the source and destination nodes is 100. The simulation results are averaged over 200 runs and presented with 95 % confidence interval. We simulate the DSR routing protocol over an ideal and contention-free channel, i.e., all the nodes within transmission range receive packet transmission correctly. MATLAB is used instead of NS2 because the intention is to compare our report-based scheme with the existing receipt-based scheme and the channel should have the same effect on the two schemes. The statistics of the simulated network demonstrate that the probability that a route has fewer than seven nodes is 0.89, the average route length is 4.21, and the network connectivity is 0.98.

3.5.2.2 Simulation Results

A. Storage Overhead

The sizes of receipts, payment reports, and *Evidences* depend on the number of intermediate nodes because the nodes' identities are attached to them. In RACE, the simulation results indicate that the *Evidence* size is $43.84 + 80i$, where i is the number of used hash chains. Table 3.12 gives the average size of receipt, report, and *Evidence* for RACE and receipt-based schemes with using 1,024-bit RSA signature scheme and SHA-1 hash function. The receipt size of Sprite is large due to attaching a signature from each end node. The receipt and *Evidence* sizes in PIS and RACE are much smaller than the receipt size in Sprite due to hashing the source and destination nodes's signatures. In RACE, the *Evidence* size is larger than the receipt size in PIS due to attaching four hash values to replace the source and destination nodes' signatures with hashing operations. For RACE and PIS, 1MB storage area can store up to 7,289.9 and 16,425 *Evidences* and receipts, respectively when one hash chain is used in RACE. Hashing the nodes' signatures can alleviate the effect of the long RSA signature tag. Although PIS requires less storage area, it needs two signatures per message: one from the source node and another from the destination node.

In RACE, the *Evidence* size depends on the number of used hash chains (i) because two hash values should be attached to the *Evidence* per hash chain. If the hash chain size is long enough, RACE can generate a fixed-size *Evidence* per route.

Table 3.12 The average size of payment receipts, *Evidences*, and reports (bytes)

	Receipt-based incentive schemes		RACE	
	Sprite	PIS	Report	*Evidence*
Upper limit	307.692	66.432	24.627	150.048
Mean	297	64	23.84	144
Lower limit	286.308	61.568	23.053	137.952

Table 3.13 The distribution of the number of used hash chains

S_{max}	Hash chain size (N)	$P(i = 1)$	$P(i = 2)$	$P(i = 3)$	$P(i > 3)$
3 m/s	30	0.48	0.24	0.11	0.17
	50	0.6	0.28	0.12	0
10 m/s	30	0.89	0.11	0	0
	50	0.99	0.01	0	0

Table 3.13 gives the distribution of the number of used hash chains. The simulation results demonstrate that more hash chains are used in low node mobility because more packets are transmitted before the route is broken. It can also be seen that the probability of using only one hash chain increases with the increase of the hash chain size (N). Properly choosing N can reduce the number of used hash chains and thus reduce the *Evidence* size. It can also save the nodes' resources because the unused hash values in a chain should not be used for other routes to secure the payment. A good N depends on the average number of transmitted packets before the route is broken, which is related to the packet transmission rate, the node speed, and the expected number of transmitted packets in the session.

Figure 3.26 shows the average *Evidences'* storage area for 1,000 *Evidences* versus the aggregation level (L). The aggregation level means that the 1,000 *Evidences* are stored in 1,000/L aggregated *Evidences*. To better clarify the effect of L on the storage area, the internal plot shows the average storage area at small aggregation level ($L \leq 20$), but the external plot shows the average storage area at large aggregation levels up to 90. The figure demonstrates that the increase of L over 10 has little impact on the storage area but increases the number of redundant *Evidences* that are submitted if an *Evidence* is requested. For example, if L is two, 500 aggregated *Evidences* are composed and each one can prove the payment for two reports, so one redundant *Evidence* is submitted if an *Evidence* is requested. Similarly, if L is 1,000, all the *Evidences* are aggregated in only one aggregated *Evidence*, and thus 999 redundant *Evidences* are submitted if an *Evidence* is requested. Figure 3.27 gives the storage area of the *Evidences* versus the number of *Evidences* without aggregation and with aggregating the *Evidences* in n_E aggregated *Evidences*. Without aggregation, the *Evidences* occupy larger storage area, and the storage area can be reduced when all the *Evidences* are aggregated in one *Evidence*.

The nodes delete the receipts after submission in receipt-based incentive schemes, but they have to store the *Evidences* until the reports are cleared in RACE. However, the nodes of MWNs are typically equipped with limited battery energy and the network is characterized with limited bandwidth. The storage area may not be the main concern, but bandwidth and energy are more scarce, i.e., reducing the amount of submitted data is more important than the size of stored data. As will be discussed in next subsection, the amount of submitted reports in RACE is much less than the amount of submitted receipts in receipt-based schemes. Moreover, the capacity of the flash memories continue to rise as per Moore's Law and their costs continue to plummet [36]. It became feasible to build cost-effective nodes with more than a gigabyte of flash memory.

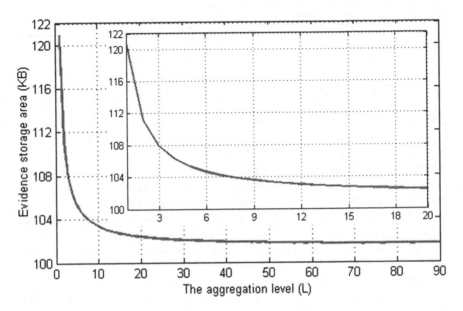

Fig. 3.26 The *Evidences'* average storage area at different aggregation levels

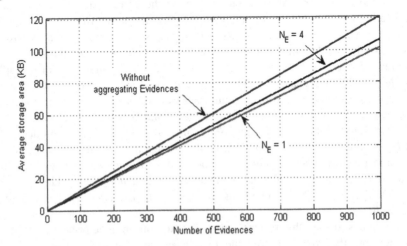

Fig. 3.27 The *Evidences'* average storage area at different number of aggregated *Evidences*

B. Payment Data Submission Overhead

The payment data submission overhead depends on the number and the size of receipts and reports. The number of receipts and reports generated in a session depends on the frequency of breaking the route between the source and destination nodes because a new receipt/report is generated when the route is broken, and therefore, the MAC layer and the simulation parameters will have the same effect on RACE and receipt-based schemes. Sprite generates a receipt per packet, but PIS and RACE generate a receipt/report per route for any number of transmitted data. From Table 3.12, RACE requires submitting only 23.84 bytes for each payment report. This amount of data is much less than those of the existing receipt-based incentive schemes because the security tokens (such as signatures) are systematically submitted in receipt-based schemes, but they are submitted only in case of cheating in RACE. Even if there are many cheaters in the network, RACE will require low payment submission overhead because cheaters are evicted from the network once they commit one cheating action. *A 512 KB data transmission is sufficient for submitting 1,765 and 8,192 receipts in Sprite and PIS, respectively, and submitting 21,992 reports in RACE.*

Table 3.14 gives the average amount of data to submit receipts and reports for 10-min data transmission at different node speed. The source and destination nodes are randomly selected, and a new route is established each time the route between the source and destination nodes is broken. It can be seen that a large amount of data are submitted in Sprite because a receipt is generated per message and the receipt size is large. Moreover, the increase of the nodes' speed increases the receipt's number because receipts are submitted for un-delivered packets. PIS requires submitting less amount of data because its receipts' size is less. The amount of data to submit the reports in RACE are much less than those of the receipt-based schemes. Table 3.14 indicates that more reports and receipts are submitted at high node mobility because the routes are more frequently broken, i.e., the source node's messages are transmitted over a larger number of routes. For RACE, a new payment report is generated when a route is broken or N (the hash chain size) messages are transmitted, but a new receipt is generated only when a route is broken in PIS. Another difference between RACE and PIS is that only one node has to submit the receipt in PIS and the other intermediate nodes submit the receipt probabilistically, but all the nodes in a route submit payment reports in RACE. However, PIS requires two signatures per packet, which consumes the nodes' resources and increases the end-to-end delay.

C. Payment Processing Overhead

Tables 3.15 and 3.16 give the processing overhead of clearing the payment of 10-min data transmission at different node speed in terms of the number of cryptographic operations, the total energy, and the processing time. We assume that Tp is a laptop with an Intel processor with 1.2 GHz and 1 GB RAM. The tables indicate that RACE does not need any cryptographic operations for clearing the payment in case of fair reports. The tables also give the overhead of verifying an *Evidence* for X messages. The simulation results demonstrate that the payment clearance overhead in RACE

Table 3.14 The average amount of submitted payment reports and receipts (KB)

Node speed		Sprite	PIS	RACE
[0, 5] m/s	Upper limit	90.511	0.549	0.2058
	Mean	87.79	0.53	0.2
	Lower limit	85.069	0.511	0.1942
[0, 20] m/s	Upper limit	91.354	0.818	0.3027
	Mean	88.18	0.788	0.293
	Lower limit	85.006	0.758	0.2833

Table 3.15 The required cryptographic operations for clearing the payment of 10-min data transmission with node speed of [0, 5] m/s

Scheme		Number of signing/ hashing / verifying operations	Energy consumption (mJ)	Processing time (ms)
Sprite	Upper limit	0 / 0 / 628	10,036	333
	Mean	0 / 0 /606	9,673	321
	Lower limit	0 /0/ 583	9,310	309
PIS	Upper limit	17.8 / 0 / 0	9,749	279
	Mean	17.1 / 0 / 0	9,345	267
	Lower limit	16.3 /0 / 0	8,941	256
	Fair reports	0 / 0 / 0	0	0
RACE	A cheating report	2 / $X + 1$ / 0	$1,093 + 0.39(X + 1)$	$31.3 + 0.029(X + 1)$

Table 3.16 The required cryptographic operations for clearing the payment of 10-min data transmission with node speed of [0, 10] m/s

Scheme		Number of signing/ hashing / verifying operations	Energy consumption (mJ)	Processing time (ms)
Sprite	Upper limit	0 / 0 / 629	10,040	333
	Mean	0 / 0 /608	9,716	322
	Lower limit	0 /0/ 588	9,393	312
PIS	Upper limit	26 / 0 / 0	14,338	410
	Mean	25 / 0 / 0	13,772	394
	Lower limit	24 /0 / 0	13,206	378
RACE	Fair reports	0 / 0 / 0	0	0
	A cheating report	2 / $X + 1$ / 0	$1,093 + 0.39 (X + 1)$	$31.3 + 0.029 (X + 1)$

is much less than the existing receipt-based payment schemes. It can also be seen that receipt-based incentive schemes requires more overhead at high node mobility because more receipts are generated due to breaking more routes, but *the nodes' speed has no effect on the payment clearance overhead in RACE if the reports are fair.* This can indicate that receipt-based incentive schemes may not be efficiently applicable when routes are frequently broken.

The low payment processing overhead can reduce the complexity and provide flexibility to the practical implementation of Tp. Moreover, since the incentive schemes use micropayment, the overhead should be much less than the payment for the effective implementation of these schemes. The payment submission and processing overhead of the receipts will be very large with taking into account the following facts: (1) the simulation results given in Tables 3.14, 3.15, and 3.16 are only for 10-min data transmission; (2) the nodes may contact Tp after a while because this connection may not be available on a regular basis and to reduce the communication overhead; and (3) once a route is broken, a new route is established with a new receipt, and thus multiple receipts may be generated per session.

RACE can significantly reduce the overhead of submitting and clearing the payment reports when cheating actions are infrequent. Widespread cheating actions are not expected in civilian applications because the common users do not have the technical knowledge to tamper with their devices. The manufacturing companies (which are limited) cannot sacrifice their reputation and face liability by making tampered devices. Moreover, cheating nodes are evicted once they commit one cheating action and changing identity is not easy or cheap, e.g., Tp can impose fees for new memberships.

References

1. C. Gentry and Z. Ramzan. Microcredits for verifiable foreign service provider metering. *Financial Cryptography, Lecture Notes in Computer Science, Springer Berlin/Heidelberg,* 3110:9–23, 2004.
2. I. Papaefstathiou and C. Manifavas. Evaluation of micropayment transaction costs. *Journal of Electronic Commerce Research,* 5(2):99–113, 2004.
3. J. Palmer and L. Eriksen. Digital newspapers explore marketing on the internet. *ACM Communications,* 42(9):33–40, 1999.
4. L. Buttyan and J. Hubaux. Stimulating cooperation in self-organizing mobile ad hoc networks. *Mobile Networks and Applications,* 8(5):579–592, Oct. 2004.
5. L. Buttyan and J. Hubaux. Enforcing service availability in mobile ad-hoc wans. *Proc. of IEEE/ACM international symposium on Mobile Ad Hoc Networking and Computing (MobiHOC'00), Boston, Massachusetts, USA,* pages 87–96, Aug. 11-2000.
6. A. Weyland. Cooperation and accounting in multi-hop cellular networks. *Ph.D. thesis, University of Bern,* Nov. 2005.
7. A. Weyland, T. Staub, and T. Braun. Comparison of motivation-based cooperation mechanisms for hybrid wireless networks. *Computer Communications,* 29:2661–2670, 2006.
8. S. Zhong, J. Chen, and R. Yang. Sprite: A simple, cheat-proof, credit based system for mobile ad-hoc networks. *Proc. of IEEE INFOCOM, San Francisco, CA, USA,* 3:1987–1997, Mar. 30 - Apr. 3 2003.
9. T. Chen and S. Zhong. Inpac: An enforceable incentive scheme for wireless networks using network coding. *Proc. of IEEE INFOCOM, San Diego, CA, USA,* Mar. 14–19 2010.
10. H. Janzadeh, K. Fayazbakhsh, M. Dehghan, and M. Fallah. A secure credit-based cooperation stimulating mechanism for manets using hash chains. *Future Generation Computer Systems,* 25(8):926–934, Sep. 2009.
11. M. Mahmoud and X. Shen. Myrpa: An incentive system with reduced receipts for multi-hop wireless networks. *Proc. of IEEE Vehicular Technology Conference (IEEE VTC'10-Fall), Ottawa, Canada,* Sept. 6–9 2010.
12. Y. Zhang, W. Lou, and Y. Fang. A secure incentive protocol for mobile ad hoc networks. *ACM Wireless Networks,* 13(5):569–582, Oct. 2007.
13. J. Pan, L. Cai, X. Shen, and J. Mark. Identity-based secure collaboration in wireless ad hoc networks. *Computer Networks (Elsevier),* 51(3):853–865, 2007.

14. M. Mahmoud and X. Shen. Fescim: Fair, efficient, and secure cooperation incentive mechanism for hybrid ad hoc networks. *IEEE Transactions on Mobile Computing (IEEE TMC)*, 11(5):753–766, May 2012.
15. S. Zhong, L. Li, Y. G. Liu, and Y. R. Yang. On designing incentive compatible routing and forwarding protocols in wireless ad-hoc networks. *Proc. of ACM MobiCom, New York, NY, USA*, pages 117–131, Aug. 2005.
16. L. Anderegg and S. Eidenbenz. Ad hoc-vcg: A trustful and cost-efficient routing protocol for mobile ad hoc networks with selfish agents. *Proc. of ACM MobiCom, San Diego, CA, USA*, Sep. 2003.
17. M. Jakobsson, J. Hubaux, and L. Buttyan. A micro-payment scheme encouraging collaboration in multi-hop cellular networks. *Proc. of the 7th Financial Cryptography (FC'03)*, 2742:15–33, Jan. 2003.
18. M. Mahmoud and X. Shen. Stimulating cooperation in multi-hop wireless networks using cheating detection system. *Proc. of IEEE INFOCOM, San Diego, California, USA*, Mar. 14–19 2010.
19. N. Salem, L. Buttyan, J. Hubaux, and M. Jakobsson. Node cooperation in hybrid ad hoc networks. *IEEE Transactions on Mobile Computing*, 5(4):365–376, Apr. 2006.
20. B. Lamparter, K. Paul, and D. Westhoff. Charging support for ad hoc stub networks. *Computer Communications*, 26(13):1504–1514, 2003.
21. M. Mahmoud and X. Shen. Pis: A practical incentive system for multi-hop wireless networks. *IEEE Transactions on Vehicular Technology*, 59(8):4012–4025, 2010.
22. M. Mahmoud and X. Shen. Dsc: Cooperation incentive mechanism for multi-hop cellular networks. *Proc. of IEEE ICC'09, Germany*, Jun. 14–18 2009.
23. M. Mahmoud and X. Shen. Esip: Secure incentive protocol with limited use of public-key cryptography for multi-hop wireless networks. *IEEE Transactions on Mobile Computing (IEEE TMC)*, 10(7):997–1010, July 2011.
24. M. Mahmoud and X. Shen. Secure cooperation incentive scheme with limited use of public key cryptography for multi-hop wireless network. *Proc. of IEEE Global Communication Conference (IEEE GLOBECOM'10), Miami, Florida, USA*, December 6–10 2010.
25. M. Mahmoud and X. Shen. A secure payment scheme with low communication and processing overhead for multihop wireless networks. *IEEE Transactions on Parallel and Distributed Systems (IEEE TPDS)*, 24(2):209–224, Feb. 2013.
26. M. Mahmoud and X. Shen. Rise: Receipt-free cooperation incentive scheme for multi-hops wireless networks. *Proc. of IEEE ICC'11, Kyoto, Japan*, June 5–9 2011.
27. A. Menzies, P. Oorschot, and S. Vanstone. Handbook of applied cryptography. *CRC Press, http://www.cacr.math.uwaterloo.ca/hac, Boca Raton, Fla.*, 1996.
28. NIST. Digital hash standard. *Federal Information Processing Standards Publication 180-1*, Apr. 1995.
29. K. Sanzgiri, D. LaFlamme, B. Dahill, B. Levine, C. Shields, and E. Belding-Royer. Authenticated routing for ad hoc networks. *IEEE Selected Areas in Communications*, 23(3):598–610, Mar. 2005.
30. Y. Hu, A. Perrig, and D. Johnson. Ariadne: A secure on-demand routing protocol for ad hoc networks. *Proc. of ACM MobiCom, Atlanta, GA, USA*, Sep. 2002.
31. National Institute of Standards and Technology (NIST). Recommendation for key management - part 1: general (revised). *Special Publication 800–57 200*, 2007.
32. J. Yoon, M. Liu, and B. Nobles. Sound mobility models. *Proc. of ACM MobiCom, San Diego, CA, USA*, Sep. 2003.
33. W. Dai. Crypto++ library 5.6.0. *http://www.cryptopp.com*, 2009.

34. N. Potlapally, S. Ravi, A. Raghunathan, and N. Jha. A study of the energy consumption characteristics of cryptographic algorithms and security protocols. *IEEE Transactions on Mobile Computing*, 5(2):128–143, Mar.-Apr. 2006.
35. Y. Zhang, W. Liu, W. Lou, and Y. Fang. Location-based compromise-tolerant security mechanisms for wireless sensor networks. *IEEE Journal Selected Areas Communication*, 24(2):247–260, 2006.
36. A. Mitra el at. High-performance low power sensor platforms featuring gigabyte scale storage. *Proc. of IEEE/ACM Measurement, Modeling, and Performance Analysis of Wireless Sensor Networks*, 2005.

Chapter 4
Secure Routing Protocols

4.1 Trust/Reputation Systems

As shown in Fig. 3.1, Tp uses a trust system to calculate trust values for each node. Tp updates the nodes' trust values with using contextual information extracted from the payment reports. The trust system performs the following two phases:

1. *Rating Calculation*: a rating is an evaluation to a node's behavior in one route. For each route, Tp calculates a rating for each node in the route.
2. *Trust Update*: a node's trust value is an overall evaluation to the node's behavior. Tp updates a node's trust values by aggregating a rating with its trust values.

4.1.1 Rating Calculation

As explained in Chap. 3, Tp can process the payment reports to extract the financial information to update the nodes' credit accounts. The reports can also be processed to extract contextual information to update the nodes' trust values. The contextual information depicts the nodes' misbehavior in terms of packet dropping and route breaking. This information is called contextual because it is not carried in the reports but extracted from the context of the reports. Tp can know whether a route is complete or broken and identify the broken link by investigating the number of messages received by each node in the route (X) and the type of the last received packet (F). A route is broken when at least one link is broken during transmitting *data*, *ACK*, or *RREP* packets. A route is complete when the source node sends all the data packets and receives all the acknowledgements from the destination node.

The underlying idea of identifying a broken link is that a packet is relayed by a node if a successor node in the route reports receiving the packet. For example if $\mathcal{N}_C$ reports receiving the data packet number 5, this entails that all the nodes between $\mathcal{N}_S$ and $\mathcal{N}_C$ have relayed the packet. As shown in Fig. 4.1a, a route is complete when

M.M.E.A. Mahmoud and X. Shen, *Security for Multi-hop Wireless Networks*, SpringerBriefs in Computer Science, DOI 10.1007/978-3-319-04603-7_4, © The Author(s) 2014 all the nodes in the route submit the same number of messages and F of one. For the broken routes, there are only four possible cases shown in Fig. 4.1b–e. In Fig. 4.1b, if the link between nodes $\mathcal{N}_A$ and $\mathcal{N}_B$ is broken during relaying the *RREP* packet, then the nodes from $\mathcal{N}_B$ to $\mathcal{N}_D$ submit $X = 0$ and $F = 0$ in their reports, but the nodes from $\mathcal{N}_S$ to $\mathcal{N}_A$ do not submit reports. Since $\mathcal{N}_A$ may drop the *RREP* packet but does not submit the report to circumvent the trust system, the system accuses the two nodes in the broken link of dropping the packet. Later, we will discuss how the trust system can precisely differentiate between the honest and the malicious nodes. In Fig. 4.1c, the link between $\mathcal{N}_B$ and $\mathcal{N}_C$ is broken during relaying the first data packet. The two nodes submit $F = 0$, but $\mathcal{N}_B$ and $\mathcal{N}_C$ submit $X = 1$ and $X = 0$, respectively. In Fig. 4.1d, the link between $\mathcal{N}_B$ and $\mathcal{N}_C$ is broken during relaying the ith data packet. $\mathcal{N}_B$ and $\mathcal{N}_C$ submit $X = i$ and $X = i - 1$, respectively. $\mathcal{N}_B$ submits $F = 0$ because the last received packet is data but $\mathcal{N}_C$ submits $F = 1$ because the last received packet is *ACK*. In Fig. 4.1e, the link between $\mathcal{N}_A$ and $\mathcal{N}_B$ is broken during relaying the *ACK* of the ith data packet. The two nodes submit $X = i$, but $\mathcal{N}_A$ and $\mathcal{N}_B$ submit $F = 0$ and $F = 1$, respectively.

A node's rating can be the probability that the node is the route breaker. So the nodes that are not in a broken link receive positive rating (0) because they cannot be the route breakers. In other words, all the nodes in a complete route receive positive ratings, and the nodes that are not in the broken link of a broken route receive positive ratings. If two nodes $\mathcal{N}_A$ and $\mathcal{N}_B$ are in a broken link, Tp cannot accuse only $\mathcal{N}_A$ of breaking the route because $\mathcal{N}_B$ may break it and compose valid *Evidence* for $X - 1$ messages instead of X messages to circumvent the trust system. In other words, we consider that $\mathcal{N}_A$ and $\mathcal{N}_B$ received X messages but relayed only $X - 1$ messages. *The rationale here is that the nodes that break the routes more frequently will be accused more and thus suffer from more trust degradation.* Moreover, an honest node can protect its trust values by not involving itself in routes with a neighbour that frequently break the routes. Moreover, the neighbours of the malicious nodes change due to the node mobility and thus the accusations are distributed instead of focusing them on few nodes. For the two nodes in a broken link, two techniques can be used to calculate their negative ratings. These techniques are called simple rating technique (*SRT*) and weighted rating technique (*WRT*) [1,2].

Simple Rating Technique: The two nodes in a broken link are equally accused of breaking the route. They receive equal negative ratings (1). *The rational of this technique is that the malicious nodes should be involved in much more broken links than the honest nodes to launch effective attacks, and therefore, they can be identified because they collect much more negative ratings.* For cases 4 to 6 in Table 3.4, all the nodes receive negative ratings (1), and thus *if an attacker manipulates its payment reports, he does not only lose credits but also receives negative ratings.* Obviously, the technique is called simple because it requires simple computations and small storage area.

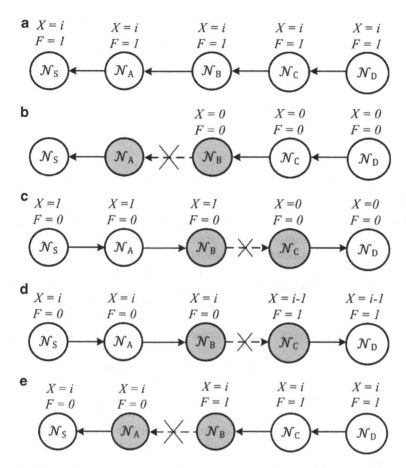

Fig. 4.1 The possible cases for complete and broken routes. (**a**) Complete route. (**b**) Broken route during relaying the *RREP* packet. (**c**) Broken route during relaying he first data packet. (**d**) Broken route during relaying the *i*th data packet. (**e**) Broken route during relaying the *i*th *ACK* packet

Weighted Rating Technique: The two nodes in a broken link receive ratings that are proportional to their rate of breaking routes in the past. If the link between nodes $\mathcal{N}_A$ and $\mathcal{N}_B$ is broken in route j, Eq. (4.1) is used to calculate the rating of $\mathcal{N}_A$ ($R_{A,j}$). The rating is the ratio of $\mathcal{N}_A$'s long-term reputation value ($R_{Lt,A}(t)$) to the summation of the two nodes' reputation values. $R_{Lt,A}(t)$ is computed from a large number of ratings and depicts the probability that the node breaks a route. By the same way, the rating of $\mathcal{N}_B$ ($R_{B,j}$) is its reputation value to the summation of the two nodes' reputation values, as given in Eq. (4.2). This is equivalent to $1 - R_{A,j}$. A rating is an evaluation to the behavior of the node in one route, but a reputation value is the overall evaluation to the node's behavior. In other words, a node's rating is the probability that the node is the route breaker, but the reputation value is the probability that it is malicious node that breaks routes intentionally.

$$R_{A,j} = \frac{R_{Lt,A}(t)}{R_{Lt,A}(t) + R_{Lt,B}(t)} \tag{4.1}$$

$$R_{B,j} = \frac{R_{Lt,B}(t)}{R_{Lt,A}(t) + R_{Lt,B}(t)} \tag{4.2}$$

As shown in Fig. 4.2, if $\mathcal{N}_A$ and $\mathcal{N}_B$ have the same reputation value, i.e., $R_{Lt,A}(t) = R_{Lt,B}(t)$, they receive equal negative ratings of 0.5, but the node having worse (higher) reputation value receives more negative rating and vice versa. *The rational of this technique is that the worse-reputation node is more likely the route breaker because it has been involved in more broken links.* The main advantage of WRT is that if honest and malicious nodes are involved in a broken link, they receive low and high negative ratings, respectively. This can improve the trust/reputation system's effectiveness because the malicious nodes' reputation values degrade much faster than those of the honest nodes do. In other words, the malicious nodes cannot cause big reduction in the honest nodes' reputation values when they are involved in broken links, but the honest nodes can cause big reduction in the malicious nodes' reputation values.

In Fig. 4.2a, $\mathcal{N}_B$ is an honest node because its reputation value is low. If $\mathcal{N}_A$ is also honest, e.g., with a reputation value between 0.05 and 0.15, the two nodes receive ratings between 0.4 and 0.6. However, if $\mathcal{N}_A$ is a malicious node, e.g., with a reputation value of 0.8, $\mathcal{N}_A$ and $\mathcal{N}_B$ receive ratings of 0.89 and 0.11, respectively. In Fig. 4.2b, $\mathcal{N}_B$ is a malicious node having a reputation value of 0.8. If $\mathcal{N}_A$ is also malicious with a reputation value close to 0.8, the ratings of the two nodes is around 0.5. In other words, a malicious node receives less negative ratings when it neighbors malicious nodes and more negative ratings when it neighbors honest nodes. Therefore, a malicious node can be identified in a shorter time when it neighbors more honest nodes because its reputation value will degrade faster. *Due to this property, if the number of honest nodes is larger than the number of malicious nodes, WRT can accelerate the degradation of the malicious nodes' reputation values,* i.e., the well-behaving majority can kick out the misbehaving minority from the network.

4.1.2   Trust/Reputation Update

Using a reputation/trust system is necessary to keep track of the nodes' long-term behavior because packets may be dropped and routes may be broken due to non-malicious reasons, such as node mobility, impaired channels, and congestion, but the high rate of route breakage is an obvious misbehavior. In order to update the nodes'

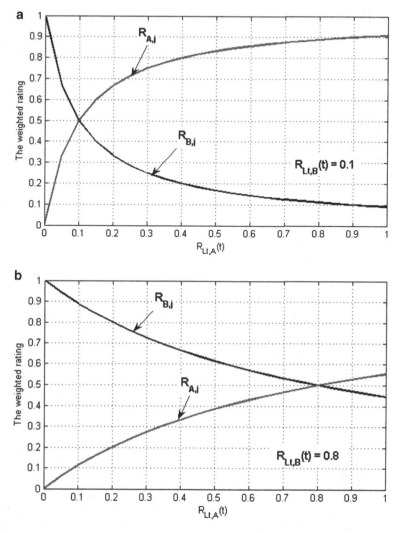

Fig. 4.2 Weighted ratings for two nodes in a broken link. (**a**) $\mathcal{N}_B$ is an honest node. (**b**) $\mathcal{N}_B$ is a malicious node

reputation/trust values, Tp aggregates the ratings with their old reputation/trust values. The reputation system stores a rating window for each node having the latest γ ratings. Figure 4.3 shows the rating window of $\mathcal{N}_A$. $R_{A,j}$ is the rating of $\mathcal{N}_A$ in route number j, and $R_{A,j} \in \{0, 1\}$ and $[0, 1]$ in *SRT* and *WRT*, respectively. After computing a new rating, the rating window is shifted to right to cancel the oldest rating ($R_{A,j}$), and the new rating is stored in the first location at right. Then, with Eq. (4.3), the short-term reputation value ($R_{St,A}(t)$) is calculated by averaging the node's latest γ ratings. This is an evaluation to the node's behavior in the latest

γ routes. Finally, with Eq. (4.4), the new long-term reputation value $(R_{Lt,A}(t))$ is calculated by aggregating $R_{St,A}(t)$ with the old long-term reputation $(R_{Lt,A}(t-1))$, where $R_{St,A}(t)$, $R_{Lt,A}(t)$ and $\alpha \in [0, 1]$. $R_{Lt,A}(t)$ expresses the probability that $\mathcal{N}_A$ is a malicious node that drop packets and break routes. $R_{Lt,A}(t)$ should be large for the malicious nodes. α is called the fading factor that determines the given weight to the nodes' past behavior. The value of α determines how fast the long-term reputation builds up and falls down, i.e., the lower value α has, the faster the old ratings are forgotten, and vice versa. To improve the effectiveness of the reputation system, α should be greater than $\alpha - 1$ because $R_{Lt,A}(t-1)$ is calculated over more routes than $R_{St,A}(t)$.

$$R_{St,A} = \frac{1}{\gamma} \sum_{k=1}^{\gamma} R_{A,k} \tag{4.3}$$

$$R_{Lt,A}(t) = \alpha R_{Lt,A}(t-1) + (1-\alpha) R_{St,A}(t) \tag{4.4}$$

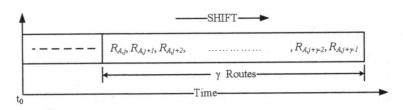

Fig. 4.3 Rating window of $\mathcal{N}_A$ having its latest γ ratings

A node's reputation value is updated by $R_{St,A}(t)$ instead of only the latest rating (good or bad) to better differentiate between the honest and the malicious nodes. By this way, *the long-term reputation values of the honest nodes degrade slower than those of the malicious nodes do because their short-term reputation values are smaller*. Similarly, *the long-term reputation values of the honest nodes improve faster than those of the malicious nodes do when they receive positive ratings because the honest nodes' short-term reputations are smaller*. Moreover, the honest nodes can filter out their negative ratings in two levels: (1) shifting the rating window forgets the node's behavior in one route, and (2) using α forgets a ratio of the node's past behavior.

The notion of trust used in this brief is defined as the degree of belief, the expectation, or the probability that a node will act in a certain way in the future based on the node's past behavior [3]. The trust values depict the nodes' reliability and competence in relaying packets, and the nodes' misbehavior. Trust values are calculated from the past behavior to predict the expected future behavior. For instance, people will not assign critical jobs to someone with a record of failure since there is a good reason to believe that he will not get the job done properly. Similarly, if a node has broken a large percentage of routes in the past, there is a strong belief that this node will break routes with high probability in the future, and thus the routing protocol should avoid it.

A trust relationship is never absolute, but it is context-dependent in the sense that a node's trust value depicts its ability to perform a specific action. For example, Alice may trust Bob to repair her computer but she may not trust Bob to repair her car. Trust is also dynamic or time-sensitive so Tp has to periodically evaluate the nodes' trustworthiness. This means that a trust value at time t may be different from its value at another time t'. In order to capture the dynamicity of trust, it should be expressed as a continuous value rather than binary or even discrete. Also, a continuous variable can represent uncertainty better than a binary variable.

Our trust system adopts multi-dimensional trust management framework in which the notion of trustworthiness is further classified into several attributes (or dimensions). Each attribute can indicate to what extent the node will conduct one specific action. We use multi-dimensional trust values instead of one trust value to precisely predict the nodes' future behavior. The trustworthiness of a node $\mathcal{N}_K$ at time t is assessed in n-dimension vector of numeric values $\tau_K(t) = [\tau_K^{(1)}(t), \tau_K^{(2)}(t), \ldots, \tau_K^{(n)}(t)]$, where $\tau_K^{(i)}(t)$ refers to the ith dimension of the trustworthiness of $\mathcal{N}_K$. Each dimension $\tau_K^{(i)}(t)$ corresponds to one action $\beta^{(i)}$. $\tau_K^{(i)}(t)$ depicts the probability that $\mathcal{N}_K$ will conduct $\beta^{(i)}$ in an appropriate manner, and thus the higher the value of $\tau_K^{(i)}(t)$ is, the more likely $\mathcal{N}_K$ will conduct $\beta^{(i)}$. $\tau_K^{(i)}(t)$ can be assigned any real value in the range of $[0, +1]$ signifying a continuous range from complete distrust (0) to complete trust ($+1$), i.e., $\tau_K^{(i)}(t) \in [0, 1], \forall i \in \{1, 2, \ldots, n\}$.

$\tau_K^{(1)}(t)$ depicts the probability that $\mathcal{N}_K$ will relay a packet successfully. From Eq. (4.5), $\tau_K^{(1)}(t)$ is the total number of packets that are relayed by $\mathcal{N}_K$ to the total number of incoming packets to be relayed by the node in the last δ routes. Obviously, if $\mathcal{N}_K$ drops a large portion of packets such as *Black-Hole* and *Gray-Hole* attackers, $\tau_K^{(1)}(t)$ will be very low. $\tau_K^{(2)}(t)$ depicts the probability that $\mathcal{N}_K$ will not break a route. Since $R_{LT,K}(t)$ is the probability that $\mathcal{N}_K$ breaks a route, $\tau_K^{(2)}(t)$ is $1 - R_{LT,K}(t)$ as given in Eq. (4.6). If $\mathcal{N}_K$ breaks a large portion of the routes, e.g., due to high mobility or misbehavior, $\tau_K^{(2)}(t)$ will be low. If $\tau_K^{(2)}(t)$ is low, that does not necessarily mean that $\tau_K^{(1)}(t)$ is low, e.g., the nodes that break many routes but after relaying a large number of packets. $\tau_K^{(3)}(t)$ depicts the probability that $\mathcal{N}_K$ can relay at least ψ packets in a route. From Eq. (4.7), $\tau_K^{(3)}(t)$ is the percentage of routes that $\mathcal{N}_K$ relayed at least ψ packets in the last δ routes. $\tau_K^{(3)}(t)$ depicts the ability of $\mathcal{N}_K$ to keep a route connected for at least ψ packets. This trust value should be low for the high-mobility nodes and the nodes that participate only in short routes. The trust values are calculated only from the last δ routes, e.g., 50–100 routes, because the recent steady behavior is a better predictor for the future behavior than behaviors observed long time ago. However, considering only the most recent behavior (very small δ) can yield a distorted picture of the nodes' behavior as few observed instances are not enough to measure the trend of the behavior.

$$\tau_K^{(1)}(t) = \alpha_1 \tau_K^{(1)}(t-1) + (1-\alpha_1) \frac{\text{No. of relayed packets in the last } \delta \text{ routes}}{\text{No. of received packets in the last } \delta \text{ routes}}$$

(4.5)

$$\tau_K^{(2)}(t) = 1 - R_{Lt,K}(t)$$

(4.6)

$$\tau_K^{(3)}(t) = \alpha_3 \tau_K^{(3)}(t-1) + (1-\alpha_3) \frac{\text{No. of routes } \mathcal{N}_K \text{ relayed at least } \psi \text{ messages}}{\delta}$$

(4.7)

$$\tau_K^{(4)}(t) = \alpha_4 \tau_K^{(4)}(t-1) + (1-\alpha_4) \frac{\text{No. of routes } \mathcal{N}_K \text{ participate in during period } \xi}{\mathcal{M}}$$

(4.8)

Since there is a stronger belief in the trust values that are computed from recent routes, $\tau_K^{(4)}(t)$ given in Eq. (4.8) depicts how $\mathcal{N}_K$ was recently active in participating in routes, or to what extent the other trust values are fresh, i.e., computed from recent routes. $\tau_K^{(4)}(t)$ is computed with using the total number of routes $\mathcal{N}_K$ participated in during the last period ξ over a normalizing factor ($\mathcal{M}$) that depicts the maximum number of routes an actively cooperative node should participate in during ξ. Note that the maximum value of $\tau_K^{(4)}(t)$ is 1, and thus $\tau_K^{(4)}(t) = \alpha_4 \tau_K^{(4)}(t-1) + (1-\alpha_4)$ if $\mathcal{M} <$ the number of routes $\mathcal{N}_K$ participated in during the last period ξ. Obviously, $\tau_K^{(4)}(t)$ decreases over time if $\mathcal{N}_K$ does not participate in routes. Since our trust system is centralized, it can determine a good value for $\mathcal{M}$ by observing the maximum number of routes the active nodes participate in during period ξ. Even if the value of $\mathcal{M}$ is not optimal, $\tau_K^{(4)}(t)$ is still valuable measure because the trust values are relative and not absolute. As will be discussed in Sect. 4.3, the routing protocol selects the nodes that have higher trust values than the other nodes and does not specify absolute trust values.

Humans are able to know each other better as time goes by and the same idea applies here. The routing protocol can trust more the nodes that spent more time in the network than the nodes that joined the network recently, because the trust system had enough time to assess their behavior. The basic idea is to use the time the nodes spent in the network as a metric in selecting the intermediate nodes.

Since Tp is not involved in the communication routes, it attaches the nodes' trust values to their certificates to enable them to prove to the other nodes that they indeed possess these trust values. From Fig. 4.4, the certificate of node $\mathcal{N}_K$ ($Cert_K$) has its identity (ID_K) and public key (K_{K-}), certificate issuing time (t_i), the time $\mathcal{N}_K$ joined the network (t_j), the certificate expiration time (t_e), the trust values ($\tau_K(t)$), and the Tp's signature on the certificate's content.

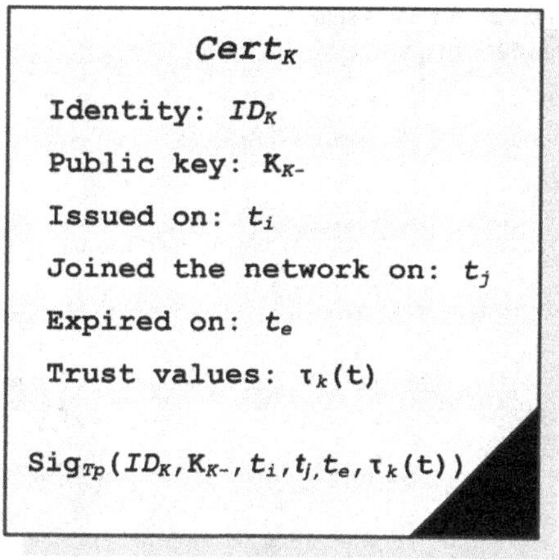

Fig. 4.4 The format of the
nodes' certificates

```
Cert_K

Identity: ID_K

Public key: K_K-

Issued on: t_i

Joined the network on: t_j

Expired on: t_e

Trust values: τ_k(t)


Sig_Tp(ID_K,K_K-,t_i,t_j,t_e,τ_k(t))
```

4.2 Identification of Malicious Nodes

A node's state is a conclusion for its behaviour based on the accumulated observations on it (or its reputation value). A node's state space includes three mutually disjoint states: (1) honest or regular node (+1), (2) suspicious or undecided (0), and (3) malicious or irrational packet dropper (-1). From Eq. (4.9), the state of $\mathcal{N}_A$ ($S_A(t)$) is honest if the node's long-term reputation value is below the honest threshold R_h; $S_A(t)$ is malicious if the node's long-term reputation value is above the malicious threshold R_m; otherwise, $S_A(t)$ is suspicious. Moreover, a node is identified as malicious when it spends ω consecutive sessions in the suspicious state because the node receives negative ratings more than the normal rate. A node is also identified as malicious when the difference between the times it spent in the honest and the suspicious states is less than β because the node receives positive ratings less than the normal rate.

The state transition diagram for a node is shown in Fig. 4.5. A suspicious node may be honest but its reputation is degraded temporarily; therefore, instead of taking a harsh reaction by characterizing this node as malicious, the reputation system keeps collecting information about the node's behavior to figure out whether its misbehavior is temporary or genuine. If a suspicious node is honest, it should be able to improve its reputation and return to the honest state, but the malicious node stays some time in the suspicious state before it is transferred to the malicious state. As shown in the figure, a node is transferred directly from the honest state to the cheating state without passing through the suspicious state when it commits a clear cheating action such as the source node in case 1 in Table 3.4.

$$S_A(t) = \begin{cases} +1, & R_{Lt,A}(t) < R_h & \text{(Honest)} \\ 0, & R_h \leq R_{Lt,A}(t) \leq R_m & \text{(Suspicious)} \\ -1, & R_{Lt,A}(t) > R_m & \text{(Malicious)} \end{cases} \qquad (4.9)$$

Fig. 4.5 A node's state
transition diagram

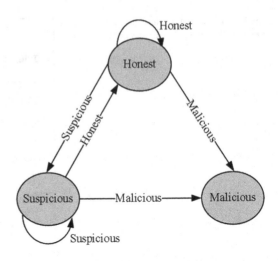

With R_m, Tp can identify the attackers that break routes more than the normal
rate, and with the threshold β, Tp can identify the attackers that break a large
number of routes in short time such as the broken nodes that misbehave after gaining
good reputation. Moreover, with ω, Tp can identify the attackers that spend long
time in the suspicious state such as *Gray-Hole* attackers. Since it is impossible to
know whether a route is broken due to non-malicious reasons or intentionally, the
attackers may break a portion of the routes that will keep their reputation values
above the reputation system's thresholds to avoid being identified as malicious.
If the thresholds are close enough to the normal rate, then the reputation system
can force the attackers to break routes at a lower rate than the system's thresholds.
This means that the system can force the smart attackers to behave in such a way
that is not a severe threat to the network proper operation. Moreover, these attackers
will have a little chance to be selected in routes. As will be explained in Sect. 4.3,
the routing protocols selects the highly trusted nodes in routes.

The threshold R_h enables the honest nodes to filter out their negative ratings
because the nodes are considered honest as long as their reputation values are less
than R_h. Tp tolerates the degradation of an honest node's reputation value up to
R_m provided that the node improves its reputation and returns from the suspicious
state to the honest state. Actually, there is an intuitive trade-off between the required
time to detect the malicious nodes and the number of honest nodes that are falsely
identified as malicious. This tradeoff can be controlled by the system thresholds. For
larger R_m, Tp tolerates more misbehavior and reduces the false accusations, but at
the expense of longer detection time. This is because the malicious nodes have to
break more routes to be identified. The *Black-Hole* attackers can be identified in
short time because their reputation values degrade fast, but it may take more time to
identify the *Gray-Hole* attackers because they build up their reputations by behaving
honestly for a while.

4.3 Reliable and Secure Routing Protocols

Although the nodes having reputation values less than R_h are considered honest, the honest nodes will have different route breaking and packet dropping rates. For example, the nodes having large hardware resources (buffer size, CPU computing power, etc) and the low-mobility nodes will have less route breaking and packet dropping rates. Trust values are used to decide which nodes to select/avoid in routing. In this subsection, we explain two routing protocols, called the *Shortest Reliable Route (SRR)* and the *Best Available Route (BAR)*, to establish the routes through the highly-trusted nodes having sufficient energy to minimize the probability of breaking the routes and maximize the packet delivery probability [4,5]. *SRR* protocol establishes the shortest route that can satisfy the source node's energy and trust requirements, but the destination node selects the best route in *BAR* protocol.

Since a trust value depicts the probability that the node conducts an action, route reliability can be computed using its nodes' trust values to give probabilistic information about the route stability and lifetime. This information can be used to establish stable routes and select proper routes that can satisfy the source nodes' requirements. Equation (4.10) gives the route reliability of a route with intermediate nodes $\mathcal{N}_W$, $\mathcal{N}_X$, $\mathcal{N}_Y$, and $\mathcal{N}_Z$, $\forall i \in \{1, 2, 3, 4\}$, to depict the probability that the action $\beta^{(i)}$ will be conducted. For example, if $i = 1$, Eq. (4.10) gives the probability that a packet will reach the destination node via the intermediate nodes $\mathcal{N}_W$, $\mathcal{N}_X$, $\mathcal{N}_Y$, and $\mathcal{N}_Z$. Similarly, $\tau_{WXYZ}^{(2)}(t)$ and $\tau_{WXYZ}^{(3)}(t)$ give the probability that the route will not be broken and the probability that at least ψ packets can be transmitted via the route, respectively.

$$\tau_{WXYZ}^{(i)}(t) = \tau_W^{(i)}(t) \times \tau_X^{(i)}(t) \times \tau_Y^{(i)}(t) \times \tau_Z^{(i)}(t) \qquad (4.10)$$

In order to clarify the importance of using the route reliability in routing decisions, numerical examples are given in Table 4.1. From cases one and two, the low-trusted node such as $\mathcal{N}_X$ in case two, nearly does not have any chance to participate in routes because it significantly degrades the route reliability. Although the nodes of cases one and three have the same trust values, the route reliability of case three is larger, which demonstrates that the shortest routes are inherently preferable. If $i = 1$, the probability of delivering a packet through the route of case four is nearly zero because the nodes have very low trust values. This case can demonstrate the importance of using trust in routing decisions. With $i = 2$ and comparing cases one and four, it is obvious that choosing a good route is important for route stability. From cases one and five, a longer route with trusted nodes is preferable than a shorter route with malicious nodes. From cases three and six, once the trust value of $\mathcal{N}_Y$ slightly decreases beyond the other nodes' trust values, the route reliability decreases, and thus the node's chance to participate in routes decreases.

Since a connection to Tp may not be available on a regular basis, the payment reports may be submitted after some time, and thus the trust values may be updated after some delay. This is acceptable because: (1) the routing protocol is sensitive to

Table 4.1 Numerical examples for route reliability

Case	$\tau_W^{(i)}(t)$	$\tau_X^{(i)}(t)$	$\tau_Y^{(i)}(t)$	$\tau_Z^{(i)}(t)$	The route reliability
1	0.8	0.8	0.8	0.8	$\tau_{WXYZ}^{(i)}(t) = 0.4096$
2	0.8	0.2	0.8	0.8	$\tau_{WXYZ}^{(i)}(t) = 0.1024$
3	0.8	0.8	0.8	–	$\tau_{WXY}^{(i)}(t) = 0.512$
4	0.2	0.2	0.2	0.2	$\tau_{WXYZ}^{(i)}(t) = 0.0016$
5	0.8	0.2	0.8	–	$\tau_{WXY}^{(i)}(t) = 0.128$
6	0.8	0.8	0.75	–	$\tau_{WXY}^{(i)}(t) = 0.48$

any degradation in the trust values; and (2) the nodes' behavior is repetitive, i.e., for a normal node the probability of breaking a route is fixed.

4.3.1   SRR Routing Protocol

In order to establish an end-to-end route to the destination node $\mathcal{N}_D$, the source node $\mathcal{N}_S$ broadcasts *Route Request Packet (RREQ)* packet and waits for *Route Reply Packet (RREP)* packet. The source node embeds its requirements in the *RREQ* packet, and the nodes that can satisfy these requirements broadcast the packet. The destination node establishes the shortest route that can satisfy the source node's requirements. The rationale of *SRR* protocol is that *the nodes that can satisfy the source node's requirements are trusted enough to act as intermediate nodes.* The protocol is useful to establish a route that averts the low-trusted nodes.

4.3.1.1   RREQ Delivery

As shown in Fig. 4.6a, the *RREQ* packet at the source node contains the packet type identifier "*RREQ*", the identities of the source and destination nodes (ID_S and ID_D), the maximum number of intermediate nodes (H_{max}), the time stamp of the route establishment (Ts), the payment-splitting ratio (Pr), the trust requirement ($T_r = [\tau^{(1)}, \tau^{(2)}, \tau^{(3)}, \tau^{(4)}]$), the energy requirement (E_r), and the source node's signature ($\{\}K_{S+}$) and certificate ($Cert_S$). K_{S+} is the private key of the source node. The source node is charged Pr of the total payment and the destination node is charged $1 - Pr$ of the total payment. H_{max} can limit the propagation area of the packet and Ts can ensure the freshness of the request. The trust requirements are the minimum trust values an intermediate node can have, e.g., if $T_r = [0.7, 0, 0, 0]$, the first trust value of the intermediate nodes should be at least 0.7 regardless of the other trust values. E_r is the minimum number of packets an intermediate node commits to relay, which depends on the node's available battery energy level. If a node breaks the route before relaying E_r data packets, its trust values will decrease. The minimum route reliability is bounded by T_r raised to the power of H_{max}.

a

$\mathcal{N}_S \rightarrow *$: RREQ, $D = (ID_S, ID_D, H_{max}, Ts, Pr, T_r, E_r), \{D\}K_{S+}, Cert_S$

$\mathcal{N}_X \rightarrow *$: RREQ, D, $\underline{ID_X}$, $\underline{\{\{D\}K_{S+}\}K_{X+}}$, $Cert_S$, $\underline{Cert_X}$

$\mathcal{N}_Y \rightarrow *$: RREQ, D, ID_X, $\underline{ID_Y}$, $\underline{\{\{\{D\}K_{S+}\}K_{X+}\}K_{Y+}}$, $Cert_S$, $Cert_X$, $\underline{Cert_Y}$

$\mathcal{N}_Z \rightarrow *$: RREQ, D, ID_X, ID_Y, $\underline{ID_Z}$, $\underline{A_Z = (\{\{\{\{D\}K_{S+}\}K_{X+}\}K_{Y+}\}K_{Z+})}$,

$\qquad\qquad\qquad\qquad\qquad\qquad\qquad\qquad\qquad Cert_S, Cert_X, Cert_Y, \underline{Cert_Z}$

b

$\mathcal{N}_D \rightarrow \mathcal{N}_Z$: RREP, $R = (ID_S, ID_X, ID_Y, ID_Z, ID_D), V_D^0, A_D = \{A_Z, V_D^0\}K_{D+}, Cert_D$

$\mathcal{N}_Z \rightarrow \mathcal{N}_Y$: RREP, R, V_D^0, A_D, $Cert_D$, $\underline{Cert_Z}$

$\mathcal{N}_Y \rightarrow \mathcal{N}_X$: RREP, R, V_D^0, A_D, $Cert_D$, $Cert_Z$, $\underline{Cert_Y}$

$\mathcal{N}_X \rightarrow \mathcal{N}_S$: RREP, R, V_D^0, A_D, $Cert_D$, $Cert_Z$, $Cert_Y$, $\underline{Cert_X}$

Fig. 4.6 The format of *RREQ* and *RREP* packets in *SRR* routing protocol. (**a**) The format of *RREQ* packet. (**b**) The format of *RREP* packet

Each intermediate node ensures that it can satisfy the source node's trust/energy requirements, the current time is within a proper range of Ts, and the number of intermediate nodes is fewer than H_{max}. It also verifies the packet's signature(s) using the public keys extracted from the nodes' certificates. These verifications are necessary to ensure that the packet is sent and relayed by legitimate nodes and the nodes can satisfy the trust requirements because their trust values are signed by Tp in their certificates. The intermediate node signs the packet's signature forming a chain of signatures signed by the nodes that broadcast the packet. This signature authenticates the intermediate node and proves that the node is the certificate holder and thus it possesses the attached trust values. The signature also enables the trust system to make sure that the intermediate node has indeed participated in the route to secure the trust calculations. Finally, as shown in Fig. 4.6a, the intermediate node broadcasts the packet after adding the signature chain and its identity and certificate. If a node receives the same *RREQ* packet from different neighboring nodes, it processes only the first packet and discards the subsequent packets. The source node's requirements cannot be achieved if it does not receive *RREP* packet within a pre-defined time period. The source node can initiate a second round of *RREQ* packet but with more flexible requirements, e.g., by increasing H_{max} and/or decreasing E_r and T_r, or revert to *BAR* protocol.

4.3.1.2 Route Selection and RREP Delivery

If there is a route that can satisfy the source node's requirements, the destination node receives at least one *RREQ* packet. The destination node composes the *RREP* packet for the route traversed by the first received *RREQ* packet, and sends it to the source node. This route is the shortest one that can satisfy the source node's requirements. $\mathcal{N}_D$ verifies the signature chain in the *RREQ* packet $(A_Z = \{\{\{\{D\}K_{S+}\}K_{X+}\}K_{Y+}\}K_{Z+})$. If the verification fails for the first arrived *RREQ* packet, $\mathcal{N}_D$ verifies the signature chain of the second packet and so on. By this way, if an intermediate node manipulates the signature chain, it cannot prevent establishing the route. As shown in Fig. 4.6b, *RREP* packet contains the packet type identifier "*RREP*", the identities of the nodes in the route (R), V_D^0, and the destination node's certificate and signature $(A_D = \{A_Z, V_D^0\}K_{D+})$. V_D^0 is the root of the hash chain created by the destination node, as discussed in Sect. 3.2. A_D can authenticate the nodes in the route with less packet space than attaching separate signatures. It also authenticates the destination node's hash chain and links it to the route, and proves to Tp that $\mathcal{N}_D$ has indeed participated in the route and agreed to pay for the packets.

As shown in Fig. 4.6b, each intermediate node verifies A_D, adds its certificate, and relays the packet. It also stores V_D^0 and A_D to be used for composing the payment *Evidence*. In Sect. 3.3 and Fig. 3.8, *Auth_Code* refers to A_Z in Fig. 4.6. From the *RREQ* packet, each intermediate node can authenticate the source node and the in-between intermediate nodes, and from the *RREP* packet, each intermediate node can authenticate the destination node and the in-between nodes. For example, in Fig. 4.6, $\mathcal{N}_Y$ can authenticate $\mathcal{N}_S$ and $\mathcal{N}_X$ from the signature chain $(\{\{D\}K_{S+}\}K_{X+})$ attached to the *RREQ*, and authenticate $\mathcal{N}_D$ and $\mathcal{N}_Z$ from A_D attached to the *RREP* packet. In order to reduce the number of verifications, $\mathcal{N}_Y$ stores the signature chain $\{\{\{D\}K_{S+}\}K_{X+}\}K_{Y+}$ and decrypts A_D until it obtains $\{\{\{D\}K_{S+}\}K_{X+}\}K_{Y+}$. By this way, each node in the route performs one signature and $2(R_L - 1)$ verifications to verify A_D and the certificates, where R_L is the number of nodes in the route including the source and destination nodes. These signature verifications is necessary to make sure that A_D is correct, and thus to ensure the *Evidence*'s integrity and secure the trust/payment systems. The source node verifies A_D and the nodes' certificates to make sure that the intermediate nodes can satisfy its trust requirements and the intended destination node was reached, then it starts data transmission. For the energy requirement, if a node cannot satisfy this requirement, the route will be broken at this node and thus its trust values degrade.

4.3.2 BAR Routing Protocol

4.3.2.1 RREQ Delivery

As shown in Fig. 4.7, the *RREQ* packet contains ID_S, ID_D, H_{max}, Ts, Pr, the number of messages it needs to send $(E_r(S))$, and the source node's certificate and signature (A_S). For the first received *RREQ* packet, an intermediate node $\mathcal{N}_X$ broadcasts the packet after attaching its identity and certificate, the number of messages it commits to relay ($E_r(X)$). Unlike *SRR* protocol, $E_r(X)$ can be fewer than $E_r(S)$. $\mathcal{N}_X$ also signs the concatenation of $E_r(X)$ and the signature received in the *RREQ* packet. $E_r(X)$ not only depends on the available battery energy level in $\mathcal{N}_X$, but also on other factors such as the cooperation strategy (the node's willingness to relaying packets) and the link quality and stability. For example if the links between $\mathcal{N}_X$ and its two neighbors in the route are unstable, it can decrease $E_r(X)$ to decrease the probability of breaking the route. The nodes are motivated to report correct energy commitments to avoid breaking the route and thus degrading their trust values.

Blind *RREQ* flooding generates few routes because each node broadcasts the packet once, which may eliminate potential better routes. To resolve this issue, *BAR* allows each node to broadcast the *RREQ* packet more than once if the route reliability or lifetime of the recently received packet is greater than those of the last broadcasted packet. The route lifetime is the minimum number of packets the intermediate nodes commit to relay, e.g., if the commitments of the intermediate nodes are $E_r(X) = 10$, $E_r(Y) = 8$, and $E_r(Z) = 17$, the route lifetime is 8 packets. For example, in Fig. 4.8, $\mathcal{N}_M$ receives the first *RREQ* packet at time t_1 with route reliability ($\tau_{AB}^{(i)}$) of 0.23. At t_2, $\mathcal{N}_M$ broadcasts the packet, where the updated route reliability is $\tau_{ABM}^{(i)} = \tau_{AB}^{(i)} \times \tau_M^{(i)} = 0.18$. At t_3, $\mathcal{N}_M$ receives the second *RREQ* packet for the same request and it discards the packet because $\tau_{NFK}^{(i)}$ is less than $\tau_{AB}^{(i)}$. At t_4, $\mathcal{N}_M$ receives *RREQ* packet with $\tau_{WXYZ}^{(i)}$ that is larger than the route reliability of the last broadcasted packet ($\tau_{AB}^{(i)}$), so it broadcasts the packet at t_5. In this example, we consider only the route reliability of one trust value for simplicity, but all the trust values can also be considered with using weighting factors. The source node can attach the weighting vector $[W_1, W_2, W_3, W_4]$ to the *RREQ* packets, where $W_1 + W_2 + W_3 + W_4 = 1$. Node $\mathcal{N}_M$ calculates the total route reliability as follows ($W_1 \times \tau_{AB}^{(1)} + W_2 \times \tau_{AB}^{(2)} + W_3 \times \tau_{AB}^{(3)} + W_4 \times \tau_{AB}^{(4)}$), and broadcasts a *RREQ* packet if its total route reliability is larger than that of the last broadcasted packet. The source node can also add some conditions to the *RREQ* packet such as the minimum time a node has been in the network.

$\mathcal{N}_S \rightarrow$ *: RREQ, $D = (ID_S, ID_D, H_{max}, Ts, Pr, E_r(S))$, $A_S = \{D\}K_{S+}, Cert_S$

$\mathcal{N}_X \rightarrow$ *: RREQ, D, **ID_X, $E_r(X)$, $A_X = \{A_S, E_r(X)\}K_{X+}$**, $Cert_S$, **$Cert_X$**

$\mathcal{N}_Y \rightarrow$ *: RREQ, D, ID_X, $E_r(X)$, **ID_Y, $E_r(Y)$, $A_Y = \{A_X, E_r(Y)\}K_{Y+}$**, $Cert_S$, $Cert_X$, **$Cert_Y$**

$\mathcal{N}_Z \rightarrow$ *: RREQ, D, ID_X, $E_r(X)$, ID_Y, $E_r(Y)$, **ID_Z, $E_r(Z)$, $A_Z = \{A_Y, E_r(Z)\}K_{Z+}$**, $Cert_S$, $Cert_X$, $Cert_Y$, **$Cert_Z$**

Fig. 4.7   The format of *RREQ* packet in the *BAR* protocol

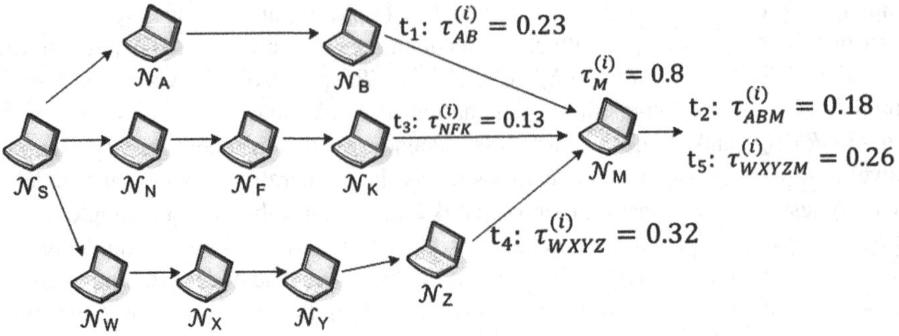

Fig. 4.8 Broadcasting the *RREQ* packet in *BAR* protocol

To reduce the number of *RREQ* broadcastings, when an intermediate node receives a *RREQ*, it introduces a *Waiting Period* to collect subsequent packets, if any, traveling through different routes. Then, it selects the most reliable route having at least lifetime of $E_r(S)$; and if this route does not exist, it selects multiple *RREQ* packets with at least total lifetime of $E_r(S)$ in such a way that reduces the *RREQ* packets' number and maximizes the reliability.

4.3.2.2 Route Selection and RREP Delivery

After receiving the first *RREQ* packet, the destination node waits for a while to receive more *RREQ* packets if there are. Then, it selects the best available route if a set of feasible routes are found. The destination node excludes the routes with very low reliability. If there are multiple routes with lifetimes at least $E_r(S)$, the destination node selects the most reliable route, otherwise, it establishes multiple routes with at least total lifetime of $E_r(S)$ in such a way that reduces the routes' number and maximizes the reliability. For example, in Fig. 4.9, the destination node receives two possible routes ($\mathcal{N}_A$, $\mathcal{N}_B$, $\mathcal{N}_C$) and ($\mathcal{N}_X$, $\mathcal{N}_Y$, $\mathcal{N}_Z$, $\mathcal{N}_W$) and chooses the most reliable one, assuming the two routes' lifetimes are more than $E_r(S)$. The destination node should not select multiple routes with common node(s) (if possible) to disallow one node to break the routes.

After selecting the best available route, the destination node sends back *RREP* packet. Composing and processing the *RREP* packet are similar to those of *SRR* protocol. The *RREP* packet of the *RREQ* packet of Fig. 4.7 should have $A_D = \{A_Z, V_D^0\}K_{D+}$ and nodes' energy commitments ($E_r(S)$, $E_r(X)$, $E_r(Y)$, $E_r(Z)$). Similar to *SRR* protocol, each node stores V_D^0, A_D, and the nodes' energy commitments for composing the *Evidence*.

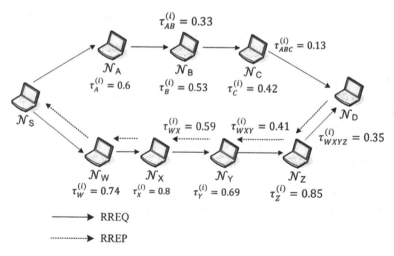

Fig. 4.9 Route selection in the *BAR* protocol

4.4 Evaluations

4.4.1 Defence Against Trust Manipulation and Irrational Route Breaking Attacks

Cooperation enforcement schemes [6–11] may not have sufficient time to precisely judge the nodes' real behavior as the period of interaction with any node may be brief due to the nodes' mobility. Our scheme can monitor the nodes over different routes and long period of time to precisely identify their behaviors. Moreover, the nodes are motivated to drop packets in cooperation enforcement schemes because packet relay consumes their resource without benefits, but packet-relay is beneficial for the nodes in our scheme to earn credits.

Reputation and trust systems are susceptible to two well known attacks called *Trust Boost* and *False Accusations*. For *Trust Boost* attack, the attackers attempt to falsely boost their trust values to avoid being identified as malicious nodes and to increase their chance to participate in routes and thus earn more credits. In *False Accusations* attack, the attackers attempt to degrade the honest nodes' trust values by falsely accusing them of breaking routes. In our trust system, the singular attackers cannot launch *Trust-Boost* attacks. For *False-Accusation* attack, the attacker has to neighbor the victim node and break the route intentionally to let *Tp* accuse its neighbor. First, neighboring the victim node is not easy due to the nodes' mobility. Second, the attacker is also accused of breaking the route and receives negative rating, which may discourage the attack. Third, frequently launching the attack will degrade the attackers' trust values and thus the effectiveness of the attack will degrade because the attackers will be less frequently selected in the routes.

This can be specifically observed in *WRT* because the attackers will receive more negative ratings than the honest nodes. Finally, falsely accusing a node of breaking a route does not guarantee that this accusation will be effective because the node can improve its trust values by participating in other routes.

Although the honest nodes may receive negative ratings when they neighbor malicious nodes, the neighbors change due to node mobility, which can distribute the negative ratings instead of concentrating them on few nodes. Moreover, since dropping the *RREQ* packets is not an abuse, an honest node can protect its reputation/trust values by not involving itself in routes with the neighbors that frequently drop packets.

Reputation/trust systems are susceptible to collusion attacks due to the nature of these systems. The impact of small-scale collusion attacks can be mitigated by categorizing ratings by identities. The system can construct *Neighbor Density Tables* (*NDTs*) for the negative and positive ratings in the rating windows. The negative rating density of $\mathcal{N}_B$ in the *NDT* of node $\mathcal{N}_A$ is the number of negative ratings that $\mathcal{N}_B$ was neighbor to $\mathcal{N}_A$ to the total number of negative ratings in the rating window of $\mathcal{N}_A$. Similarly, the positive rating density of $\mathcal{N}_B$ in the *NDT* of $\mathcal{N}_A$ is the number of positive ratings that $\mathcal{N}_B$ was neighbor to $\mathcal{N}_A$ to the total number of positive ratings in the rating window of $\mathcal{N}_A$. In other words, *NDT* can indicate the frequency that $\mathcal{N}_B$ caused positive and negative ratings to $\mathcal{N}_A$. Obviously, in small-scale collusion attacks, the colluders have much higher densities than those of other nodes.

Considering the *NDT* when deciding a node's state can improve the trust system's robustness. For example, in *Trust-Boost* attacks, few nodes contribute too much to a node's positive ratings and the node's reputation becomes bad when excluding these false ratings. Similarly, in *False-Accusation* attacks, few nodes contribute too much to a node's negative ratings and the node's reputation becomes good when excluding these false ratings. Therefore, the *NDT* can prevent a small number of colluders from falsely improving their reputation values or evicting honest nodes from the network, and thus using *NDT* forces the attackers to collude with a large number of nodes, which is hard in civilian and large-scale networks [12]. Certainly, if the *NDT*'s node densities are flat or dominated by a large number of nodes, then the reputation system can have stronger belief in the node's ratings. Several measures can be taken to improve the robustness against large-scale collusion attacks. In order to make fabricating routes by colluding nodes to boost their trust expensive, clearance fee can be imposed to clear the payment of a route. If colluders tamper their payment reports to accuse a victim, they lose credits and defame their reputation such as cases 4 to 6 in Table 3.4.

Thwarting *Trust Boost* and *False Accusation* attacks launched by large-scale colluding nodes has been studied in Web-related trust systems [13]. It is shown that *False Accusation* attacks can be avoided by concealing the real identities of the intermediate nodes. In [14], statistical filtering algorithms have been proposed to filter out the false accusations by excluding or giving low weight to the presumed unfair ratings based on analyzing the rating values. The assumption is that unfair ratings can be recognized from their statistical properties. Since eBay [15] charges a fee for each transaction, *Trust Boost* attack would be expensive. Fortunately, these techniques can be used with our trust system more effectively because it is difficult to obtain multiple identities comparing to Web applications that usually provide free access to the users via simple registration process.

Equation (4.11) gives the probability that a node receives at least a ratio of R_m negative ratings in γ routes ($P_i(\gamma)$), where P is the probability of receiving a negative rating in a route. Obviously, P should be much larger for the malicious nodes than the honest nodes because they break the routes more frequently. Figure 4.10 shows that if R_m is chosen in the range $[0.35, 0.55]$, the reputation system can perfectly differentiate between the malicious and the honest nodes. However, if R_m is low, e.g. $[0, 0.35)$, some honest nodes may be falsely identified as malicious, and if R_m is too tolerant, e.g., $(0.55, 1]$, some malicious nodes may not be identified. Thus, R_m can control the tradeoff between the false accusation probability and the malicious nodes' detection probability. The increase of P increases $P_i(\gamma)$ for the same R_m, and thus some honest nodes may be falsely identified as malicious if R_m does not have enough tolerance.

$$P_i(\gamma) = \sum_{k=\gamma R_m}^{\gamma} \binom{\gamma}{k} P^k (1-P)^{\gamma-k} \tag{4.11}$$

From Fig. 4.11, the honest and the malicious nodes can be identified more precisely with increasing the rating window size (γ) because $P_i(\gamma)$ is less for the honest nodes and more for the malicious nodes. This can reduce the number of honest nodes that falsely identified as malicious and the number of malicious nodes that are not detected. For example, if R_m is in the range $[0.3, 0.7]$, some honest

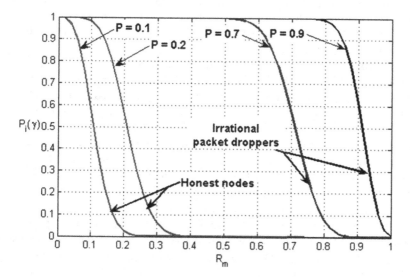

Fig. 4.10 The effect of R_m on the reputation system's effectiveness when $\gamma = 50$

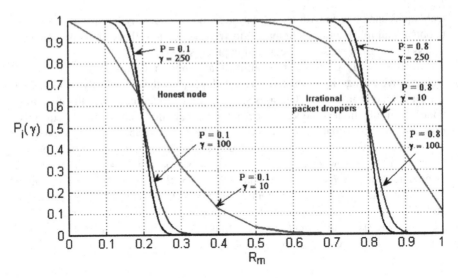

Fig. 4.11 The effect of γ on the reputation system's effectiveness

nodes may be falsely identified as malicious and some malicious nodes may not be detected when $\gamma = 10$, but the reputation system can perfectly identify the nodes' behaviors when γ is 100 or 250. However, this does not mean that the system have to wait long time until the nodes participate in γ routes for identifying the malicious nodes because a node'sА rating window has initial values when the node first joins the network. From Fig. 4.12, the aggressive attackers that break a large portion of routes, i.e., with large P, can be identified after they participate in a few number of routes (or in shorter time). For example, the probabilities to identify the malicious nodes after participating in 20 routes are 0.87, 0.92, and 0.96 when P is 0.6, 0.63, and 0.66 respectively.

A network simulator is used to evaluate the effectiveness of *SRT* and *WRT* techniques in identifying the malicious nodes. Thirty-five mobile nodes with 125 m radio transmission range are randomly deployed in a square cell of 1,000 m by 1,000 m. We adopt the modified random waypoint model [16] to emulate the nodes' mobility. The nodes' speed is uniformly distributed in the range [0, 10] m/s and the pause time is 10 s. The constant-bit-rate traffic source is implemented in each node and the source and destination pairs are randomly selected. The *DSR* routing protocol [17] is simulated over the distributed coordination function of the IEEE 802.11 medium access control protocol. Ts, a node's identity (ID_A), and the number of messages (X) are 5, 4, and 2 bytes, respectively. 300 sessions are held in each updating time, packet transmission rate is 0.5 packets/s, and 25 packets are transmitted in each session. A new route is established when the session route is broken. To determine the thresholds, we run a training phase with assuming that all the nodes are honest to measure the expected and tolerable rate of breaking the routes. The parameters R_h, ω, β, γ and α are 0.19, 100, 100, 50, and 0.78,

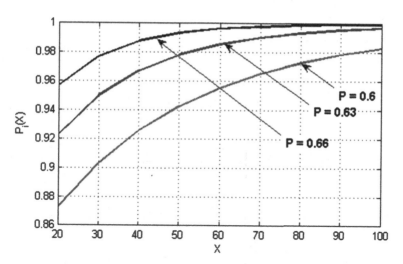

Fig. 4.12 The effect of P on $P_i(X)$ when $R_m = 0.5$

respectively. The initial rating window is the repeat of the pattern "00001", and thus the nodes' initial reputation values and states are 0.2 and honest.

In Table 4.2, the number of false-positive nodes is the average number of the honest nodes that are falsely identified as malicious, and the detection time is the average number of updating times for identifying all the malicious nodes. The attack strength $1 : X$ means that the attackers break one route intentionally and behave normally in $X - 1$ routes to circumvent the reputation system. However, in order to launch effective attacks, the attacker has to break a large percentage of the routes, i.e., X should be small. When $X = 1$, the attackers launch *Black-Hole* attacks by dropping all the packets and breaking all the routes they participate in, otherwise they launch *Gray-Hole* attacks.

The simulation results demonstrate that R_m can control the intuitive tradeoff between the detection time and the number of false-positive nodes. Less tolerance to the negative ratings ($R_m = 0.35$) can shorten the detection time but increases the number of false-positive nodes. This tradeoff is sharper in *SRT* than *WRT* because the honest nodes collect more negative ratings. It takes longer to identify the nodes that misbehave less frequently such as the nodes with 1:2 attacking strength because they lose their reputation values slowly, but they also harm the network less. The increase of the ratio of attackers increases the number of false-positive nodes because the honest nodes collect more negative ratings due to neighboring more malicious nodes, and the victims could not improve their reputation values with the rate that can keep them in the honest state. Nevertheless, for $R_m = 0.35$ and $X = 1$, when 42.86% of the nodes behave maliciously, almost no node is falsely accused in *WRT*, but around 13.25 nodes (37.8%) are falsely accused in *SRT*. This is because the honest nodes receive less negative ratings in *WRT*, i.e., *WRT* can precisely identify the malicious nodes because it can better filter out the negative ratings of the honest nodes.

Table 4.2 Simulation results

Attack strength	R_m	Attackers' ratio	Detection time (in updating times)		Number of false-positive nodes at $\gamma = 50$		Number of false-positive nodes at $\gamma = 15$	
			SRT	WRT	SRT	WRT	SRT	WRT
1:1	0.35	5%	1	1.85	2.8	0	8.85	0.16
		42.86%	1.9	4.4	13.35	0.1	16.17	0.56
	0.6	5%	2.25	14	0	0	0.9	0
		42.86%	5.3	98	0.85	0	2.85	0
1:2	0.35	5%	1.15	10.95	3.6	0	11.35	0.72
		42.86%	2	23.65	10	0	12.6	2.24
	0.6	5%	38.8	102.85	0	0	1.75	0
		42.86%	95.75	109.8	0.2	0	5.95	0

The number of false-positive nodes can be reduced in *SRT* by increasing R_m, e.g., for $X = 1$ and 42.86% of the nodes are malicious, increasing Rm from 0.35 to 0.6 reduces the number of false-positive nodes from 13.35 (37.8%) to 2.8 (8%). However, the increase of R_m means that the smart attackers can break more routes with keeping their reputation values above the system thresholds and the detection time increases. R_m can be less in *WRT*, e.g., $R_m = 0.35$, for small detection time and low number of false-positive nodes. This is because the malicious nodes collect more negative ratings and the honest nodes collect less negative ratings comparing to *SRT*. Moreover, the simulation results demonstrate that the increase of the number of attackers increases the detection time because some malicious nodes may not participate in any routes during an updating time and the malicious nodes receive less negative ratings due to neighboring more malicious nodes in *WRT*. In addition, the number of false-positive nodes increases with reducing γ, which confirms the observation shown in Fig. 4.11.

Since the reputation system's thresholds have direct impact on the system's effectiveness, our centralized reputation system can compute the thresholds from the nodes' reputation values and periodically tune them. For example, if the reputation values of the majority of the nodes are less than 0.3, R_h can be decided as 0.3 assuming that the majority of the nodes behave honestly. Moreover, since the nodes contact Tp over discrete times, the detection and eviction times can be reduced with issuing shorter-lifetime certificates to the nodes having bad reputation. Moreover, investigating the reputation's rate of change ($\frac{dR_{LT,i}(t)}{dt}$) can reduce the detection time of some attackers, e.g., the reputation's rate of change is higher for the *Gray-Hole* attackers than those of the honest nodes.

The reputation system aims to identify the malicious nodes that break a large ratio of routes, but *the trust system aims to identify the good nodes* to be selected in routing. The rationale behind using the trust system is that no need to wait until a node's reputation value degrades to a low value to be excluded from routing because they will break many routes. *Once a node's trust values fall behind those of the majority of the nodes, the node is almost excluded from routing.* In order to alleviate the tradeoff between missed detections and false accusations, R_m can be large to reduce the number false positive nodes and the routing protocol can exclude the nodes having low trust values.

Even if the *Gray-Hole* attackers drop a low ratio of the packets, they have little chance to participate in routes. This is because the honest nodes that do not drop packets intentionally will have larger trust values and will be selected by the routing protocol. Greedy nodes may overload themselves by participating in many routes simultaneously to collect more credits. This behavior will create congested nodes that drop packets once their buffers are full. Selecting the highly trusted nodes in routing can discourage this behavior because it reduces the nodes' trust values.

Our scheme uses multiple levels of trust as follows. (1) the micropayment stimulates the nodes to behave well to earn credits; (2) using certificates signed by a trusted party enables the nodes to ensure that the other nodes are members in the network; and (3) the trust values computed by the trust system is a trust metric that is based on monitoring the nodes' behaviors. Continuously evaluating the nodes' trust values is necessary to track the change in the nodes' behavior. Some nodes may change their behavior after gaining high trust values due to faulty hardware or software or malicious action.

Since the behavior of the newly joined nodes is unknown, these nodes will not be involved in a large number of routes until they build up good trust by behaving well in the routes they participate in. The nodes with good past behavior are more trusted than those with unknown behavior. The newly joined nodes will be selected when the source and destination nodes have limited options, or when they report high energy capability, so they will build up their trust values slowly. This coincides with the meaning of trust, i.e., a node cannot be trusted before showing a clear trustful behavior. Routes can be broken due to non-malicious reasons such as temporary bad channel or network congestion. The effect of this on the nodes' trust values can be neutralized by computing the trust values based on the nodes' behavior in a number of routes. It is not reasonable to assume that some nodes will suffer from the non-malicious packet drop more than others over a number of routes.

If a malicious node can create several fake identities, the trust system may suffer from *Sybil* attack [18] and the attackers can launch effective attacks. For the *Newcomer* attack [19], the attackers can easily remove their bad history when their trust values degrade by changing their identities and registering as new users. This attack can improve the effectiveness of *False Accusation* attacks. In order to make the *Newcomer* and *Sybil* attacks expensive and ineffective, T_p can impose some fees for registering a new user or changing an identity. Moreover, the nodes' initial trust values should not be high when they first join the network, and the old nodes that have spent long time in the network should be trusted more.

For *On-off* attack, the attackers take advantage of the fact that bad behaviors can be removed with good behaviors by behaving well until gaining high trust value and then they alternate their behaviors between misbehaving and well-behaving with keeping their trust values high [20]. This attack is possible when the increase in the trust value when a good behavior is observed equals to the decrease in the trust when a bad behavior is observed. Our trust system is not vulnerable to this attack because the trust values are computed based on the ratio of the good observations in the last ω sessions.

Each rating in *SRT* and *WRT* can be stored in one and seven bits, respectively, and thus, a rating window for 320 ratings requires a storage area of 40 and 280 Bytes, respectively. Moreover, the storage area can be significantly reduced by making the size of the rating window variable. The rating windows can be short for the good-reputation nodes and long for the bad-reputation nodes, e.g., suspicious nodes, to better judge their behavior. In addition, the overhead can be significantly reduced by running the reputation system only on-demand when Tp notices that the routes are broken more than the normal rate, and the system can be run only for the suspected nodes that are frequently involved in broken links.

4.4.2 Evaluations of SRT and WRT Routing Protocols

The objectives of using trust in route selection are as follows: (1) to foster trust among the nodes by making knowledge about the nodes' past behaviors available; (2) to encourage the nodes to provide high packet-relay success ratio and report correct battery energy level by giving more preference to the highly trusted nodes in route selection; and (3) to punish the nodes that provide low packet-relay success ratio because any loss of trust means loss of potential earnings.

To secure the routing protocol, our protocols can satisfy the following requirements: (1) valid routing packets (*RREQ* and *RREP*) cannot be fabricated; (2) invalid routing packets cannot be propagated in the network; (3) valid routing packets cannot be altered in transit without detection, except according to the normal functionality of the protocol; (4) routing loops cannot be formed by malicious action; (5) stale routing packets are not accepted by the nodes or propagated in the network; and (6) The packets of unauthorized nodes are not accepted by the authorized nodes.

For *Unauthorized Participation* attack, the nodes that are not members of the network should not act as source, intermediate, or destination nodes. For *Spoofed RREQ* attack, since only the source node can sign with its private key, the attackers cannot spoof other nodes to establish routes under their names. *RREQ* packets will not be accepted by the nodes if their signatures are invalid. This can thwart many external attacks launched by unauthorized nodes, such as *RREQ Flooding* attack. For *Alteration of Routing Packets* attack, an adversary attempt to manipulate the *RREQ* packets, e.g., to provide wrong data to increase its chance to be involved in routes to earn more credits or break the routes. Since routing packets are signed, manipulating the routing information, the trust/energy requirements, and the maximum number of hops of an in-transit packet would be detected by the intermediate nodes along the route, and the altered packet would be subsequently discarded. This attack cannot prevent establishing a route because flooding the *RREQ* packets enables the destination nodes to receive multiple routes. Also, dropping the *RREP* packets cannot prevent establishing the route because the destination node can send a new *RREP* packet for a different route if the data transmission does not start on a route.

In route establishment, the nodes that report incorrect trust values can be detected because the trust values are signed by Tp. The nodes cannot manipulate their trust values because they cannot forge the Tp's signature. For *Destination Node Impersonation* attack, the attacker attempts to send *RREP* packet to let the source node believe that it communicates with the destination node. This is infeasible in our routing protocols because the destination nodes sign the *RREP* packets to ensure that only the destination node can respond to the *RREQ* packet. For the *RREQ Flooding* attack launched by internal attackers, since the source nodes sign the *RREQ* packets, the attackers can be identified in an undeniable way. The network nodes can ignore a node's packets when it sends a large number of *RREQ* packets in a short time. For *Route Lengthening* attack, elongating a route by inserting non-existing nodes to the

RREQ packet requires signing the packet with the private keys of these nodes. It also decreases the chance of selecting the route because the route reliability decreases.

A route may be broken because of link failure, node failure, or node mobility. Establishing routes through stable links has been extensively studied in the literature and the proposed techniques can be integrated with our routing protocols. The link status can be considered when the nodes compute the number of messages they can relay. Our routing protocols mainly aims to reduce route breakage due to node failure and malicious action. Our protocols aim to increase the chance that packets will reach the destination nodes by relaying them by the highly-trusted nodes having sufficient energy. A node's trust values depict its failure probability. The trust values are low when the node drops packets intentionally or reports incorrect battery energy level. The trust values are also low when a node's hardware/software frequently fails or has high mobility.

The *Route Breaking* and *Packet Dropping* attacks degrade the network performance significantly. The average throughput degrades by 16 to 32 % if 10 to 40 % of the nodes drop packets, and the end-to-end packet delay increases linearly with the increase of the number of attackers [6, 21]. Moreover, since a new payment report and *Evidence* are generated each time a route is broken and re-established, breaking routes increases the payment reports submission and clearance overhead and wastes the consumed energy and bandwidth in transmitting the packets from the source node to the attacker. Breaking routes exhausts the nodes' resources in re-establishing the broken routes. It also wastes the end nodes' credits because they pay for the un-delivered messages.

In *MWNs*, a packet cannot reach to its destination if any intermediate node drops the packet, and thus the packet delivery ratio decreases with the growth of the number of attackers. Equation (4.12) gives the probability of breaking a route with R_L nodes (or $n = R_L - 2$ intermediate nodes) due to malicious action. P_m is the ratio of the malicious nodes, which is equivalent to the probability that an intermediate node is malicious. The packet delivery ratio of a route with R_L nodes ($PDR(R_L)$) is the number of received data packets to the total number of packets sent in the route. In Eq. (4.13), $PDR(R_L)$ and $PDR_0(R_L)$ are the average packet delivery ratio with and without malicious nodes, respectively.

$$P_b(R_L) = 1 - (1 - P_m)^{R_L - 2} \tag{4.12}$$

$$PDR(R_L) = PDR_0(R_L) \times (1 - P_b(R_L)) \tag{4.13}$$

Figure 4.13 shows that a low ratio of malicious nodes such as 20% can reduce the packet delivery ratio by 74% and 60% for routes with eight and six nodes, respectively. Moreover, the increase of R_L or P_m increases route breaking probability and thus degrades the packet delivery ratio.

We simulate a multihop wireless network by randomly deploying 55 nodes in an area of 1000×1000 m. n is the number of nodes having low and medium trust values. The number of nodes having high trust values is $55 - n$ and their trust values are uniformly distributed in $[0.8, 1)$. The number of nodes having low trust values is

$\lfloor 0.67 \times n \rfloor$ and their trust values are uniformly distributed in $[0, 0.3)$. The number of nodes having medium trust value is $\lceil 0.33 \times n \rceil$ and their trust values are uniformly distributed in $[0.3, 0.8)$. A node with a trust value of 0.6 breaks routes with the probability of $1 - 0.6 = 0.4$. By this way, the trust values can be used to simulate the variety in the nodes' resources and the malicious actions.

The radio transmission range is 125 m, and all the nodes start the simulation with the same initial energy that is sufficient for relaying 100 messages. The data packet size is 512K bytes, H_{max} is eight. In *SRR* protocol, the trust requirement (T_r) is 0.1 and 0.6, and the energy requirement is the number of messages the source node needs to send. In order to simulate the cryptographic operations required by the routing protocols, the processing times given in Table 3.5 is used in the simulation. The given results are averaged over 100 simulation runs and presented with 95% confidence interval. In each run, 30 communication sessions with randomly chosen source and destination pairs are established. The route is re-established if it is broken before sending 12 messages. The constant-bit-rate traffic model is employed for all the connections. To simulate the node mobility, we use the random waypoint mobility model. The pause time is 2 s and the nodes' speed is uniformly distributed in [0, 1] m/s.

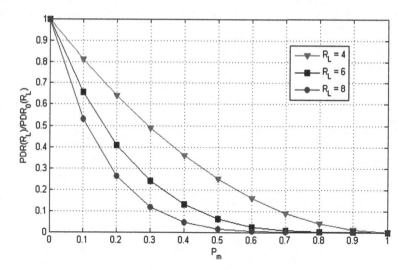

Fig. 4.13 The expected drop in the packet delivery ratio due to the malicious nodes

The packet delivery ratio is the total number of data packets received by the destination nodes to the total number of data packets sent by the source nodes. From Fig. 4.14, *SRR* and *WRT* protocols outperform the dynamic source routing (*DSR*) protocol [22] in the packet delivery ratio. *DSR* enables the low-trusted nodes and the nodes having low energy to repeatedly participate in routes and break them because it randomly chooses the intermediate nodes. Conversely, *SRR* and *WRT* protocols establish more stable routes by selecting reliable intermediate nodes, and therefore, it can deliver packets more successfully. Although *DSR* re-establishes the routes each time they are broken, the new routes may still have low-trusted nodes and thus fail again. When we compare *SRR* and *WRT* protocols with *DSR*, we actually compare between two strategies: informed routing decisions and randomly selecting intermediate nodes. *DSR* randomly selects intermediate node, but *SRR* and *WRT* protocols make informed routing decisions by selecting the nodes that behaved well in the past and have enough energy.

It can be seen that the packet delivery ratio of *DSR* significantly degrades as the number of low-trust nodes increases due to involving these nodes in routes more frequently. For *SRR*, the increase of T_r can increase the packet delivery ratio due to selecting more trusted nodes, but the probability of establishing routes decreases, as will be discussed later. When $T_r = 0.1$, the increase of n decreases the packet delivery ratio because more low-trust nodes participate in the routes. However, the reduction in the packet delivery ratio when $T_r = 0.6$ is mainly due to the messages the source nodes could not send because they did not find routes with this trust requirement. In *BAR*, the increase of the low-trust nodes has little effect on the packet delivery ratio because it can avert these nodes and select nodes with good trust values and sufficient energy. Moreover, it can be seen that when the number of low-trust nodes is few (n is small), the packet delivery ratio in *SRR* and *BAR* are larger than that of *DSR*, because they can select the nodes having sufficient energy.

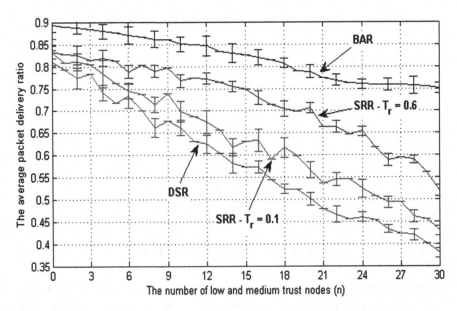

Fig. 4.14 *SRR* and *WRT* protocols can improve the packet delivery ratio due to selecting good intermediate nodes

Figure 4.15 shows the number of *RREQ* broadcasts in *SRR* and *BAR* protocols to this of the *DSR* at different values of n. The *Waiting Period* at each node is 20 ms in *BAR*. It can be seen that the normalized number of broadcasts in *SRR* is always less than one because the nodes that cannot satisfy the energy or trust requirements do not broadcast the *RREQ* packets. When $T_r = 0.6$, the number of broadcasts is less because more nodes cannot satisfy the trust requirements and thus do not broadcast *RREQ* packets. For *BAR*, the normalized number of broadcasts is always above one because the nodes may broadcast a *RREQ* packet more than once, but in *DSR*, each node broadcasts each *RREQ* packet at most once.

In Fig. 4.16, the call acceptance ratio is the ratio of times a route is established after sending a *RREQ* packet. We can see that the call acceptance ratio in *BAR* nearly does not depend on n. However, the increase of n decreases the call acceptance ratio in *SRR* because more nodes cannot satisfy the trust requirement, and thus more routes cannot be established. When $T_r = 0.6$, the call acceptance ratio significantly decreases with the increase of n because more nodes cannot satisfy the trust requirement.

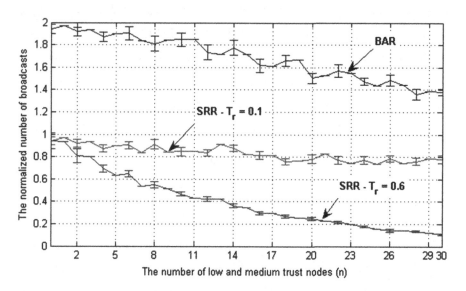

Fig. 4.15 *SRR* generates fewer *RREQ* broadcasts because the nodes that cannot satisfy the source node's requirements do not broadcast the packets

In Fig. 4.17, the normalized route lifetime is the average route lifetime in *SRR* and *BAR* protocols to that of *DSR*. The route lifetime is the number of packets sent in one route before it is broken. Route lifetime is a good measure for route stability. Since the normalized route lifetime is always more than one, *SRR* and *BAR* protocols can establish more stable routes comparing to *DSR*. When $n > 12$, *SRR* with $T_r = 0.6$ may establish more stable routes but as indicated in Fig. 4.16, the likelihood of establishing a route decreases as n increases.

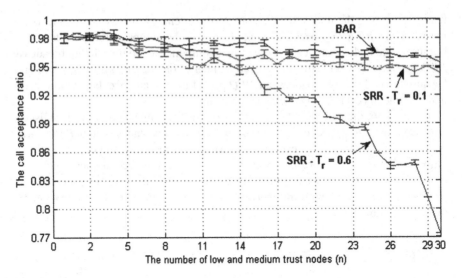

Fig. 4.16 Routes cannot be established if the source node's trust requirement is not properly determined in *SRR*

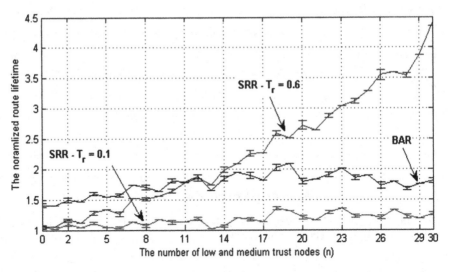

Fig. 4.17 The route lifetime in our protocols is more than that in *DSR* because of establishing more stable routes

References

1. M. Mahmoud and X. Shen. An integrated stimulation and punishment mechanism for thwarting packet dropping attack in multi-hop wireless networks. *IEEE Transactions on Vehicular Technology (IEEE TVT)*, 60(8):3947–3962, Oct. 2011.

2. M. Mahmoud and X. Shen. Credit-based mechanism protecting multi-hop wireless networks from rational and irrational packet drop. *Proc. of IEEE Global Communication Conference (IEEE GLOBECOM'10), Miami, Florida, USA,* Dec. 6–10 2010.

3. H. Li and M. Singhal. Trust management in distributed systems. *IEEE Computers,* 40(2):45–53, Feb. 2007.

4. M. Mahmoud, X. Lin, and X. Shen. Secure and reliable routing protocols for heterogeneous multihop wireless networks. *IEEE Transactions on Parallel and Distributed Systems (IEEE TPDS),* to appear.

5. M. Mahmoud and X. Shen. Trust-based and energy-aware incentive routing protocol for multi-hop wireless networks. *Proc. of IEEE International Conference on Communications (IEEE ICC'11), Kyoto, Japan,* June 5–9 2011.

6. S. Marti, T. Giuli, K. Lai, and M. Baker. Mitigating routing misbehavior in mobile ad hoc networks. *Proc. of ACM International Conference on Mobile Computing and Networking (MobiCom'00), Boston, Massachusetts, USA,* pages 255–265, Aug. 6–11 2000.

7. S. Bansal and M. Baker. Observation-based cooperation enforcement in ad-hoc networks. *Techical Report, Computer Science Department, Stanford University, CA, USA,* Jul. 2003.

8. S. Buchegger and J. Boudec. Performance analysis of the confidant protocol: cooperation of nodes - fairness in distributed ad hoc networks. *Proc. of IEEE/ACM MOBIHOC'02, Switzerland,* pages 226–236, Jun. 9–11 2002.

9. P. Michiardi and R. Molva. Core: A collaborative reputation mechanism to enforce node cooperation in mobile ad hoc networks. *Proc. of IFIP CMS'02, Portoroz, Slovenia,* pages 107–121, Sep. 26–27 2002.

10. Q. He, D. Wu, and P. Khosla. A secure incentive architecture for ad-hoc networks. *Wireless Communications and Mobile Computing,* 6(3):333–346, May 2006.

11. K. Liu, J. Deng, and K. Balakrishnan. An acknowledgement-based approach for the detection of routing misbehavior in manets. *IEEE Transactions on Mobile Computin,* 6(5), May 2007.

12. C. Cachin, K. Kursawe, A. Lysyanskaya, and R. Strobl. Asynchronous verifiable secret sharing and proactive cryptosystems. *Proc. ACM Conference on Computer and Communications Security, CCS02,* pages 88–97, 2002.

13. A. Jøsang, R. Ismail, and C. Boyd. A survey of trust and reputation systems for online service provision. *Decision Support Systems,* 43(2):618–644, 2007.

14. A. Withby A. Jøsang and J. Indulska. Filtering out unfair ratings in bayesian reputation systems. *The Icfain Journal of Management Research,* 4(2):48–64, 2005.

15. P. Resnick and R. Zeckhauser. Trust among strangers in internet transactions: Empirical analysis of ebay's reputation system. *Proc. of NBER workshop on empirical studies of electronic commerce,* 2000.

16. J. Yoon, M. Liu, and B. Nobles. Sound mobility models. *Proc. of ACM MobiCom, San Diego, CA, USA,* Sep. 2003.

17. D. Johnson and D. Maltz. Dynamic source routing in ad hoc wireless networks. *Mobile Computing, Chapter 5, Kluwer Academic Publishers,* pages 153–181, 1996.

18. J. Newsome, E. Shi, D. Song, and A. Perrig. The sybil attack in sensor networks: Analysis and defenses. *Proc. of the third International Symposium on Information Processing in Sensor Networks (IPSN),* 2004.

19. P. Resnick, R. Zeckhauser, E. Friedman, and K. Kuwabara. Reputation systems. *Communications of the ACM,* 43(12):45–48, 2000.

20. N. Bhalaji and A. Shanmugam. Reliable routing against selective packet drop attack in dsr based manet. *Journal of Software,* 4(6):536–543, Aug. 2009.

21. P. Michiardi and R. Molva. Simulation-based analysis of security exposures in mobile ad hoc networks. *Proc. of European Wireless Conference, Florence, Italy,* Feb. 25–28 2002.

22. D. Johnson, D. Maltz, and Y. Hu. The dynamic source routing protocol for mobile ad hoc networks (dsr). *technical report, IETF MANET Working Group,* Feb. 2007.

Chapter 5
Conclusions and Future Directions

In this chapter, we summarize the main ideas and concepts presented in this brief and highlight future research directions.

5.1 Conclusions

Multi-hop wireless networks have been originally proposed for military and disaster recovery applications. In this brief, we have discussed several security issues that should be resolved before considering the use of *MWNs* in civilian applications.

First, a fair and efficient incentive scheme has been presented. Micropayment is used to reward the nodes that relay others' packets and charge those that send packets. Using Micropayment in *MWNs* can stimulate the selfish nodes to relay others' packets, enforce fairness, regulate packet transmission, and discourage *Bogus Packet Overloading* attack. Micropayment systems have been originally designed for Internet-based applications. For the efficient and effective use of these systems in *MWNs*, we have introduced a payment model that is specifically tailored to cooperation stimulation in *MWNs*. The scheme adopts a fair charging policy by charging both the source and destination nodes when both of them are interested in the communication. In order to secure the payment with minimum overhead, we have introduced a communication protocol that can secure the payment with limited use of the public key cryptography. The public-key cryptographic operations are required only for the first packet, and they are replaced with the efficient hashing operations in the next packets. Therefore, for a series of packets, the large overhead of the first packet vanishes and the overall overhead converges to that of the lightweight hashing operations.

Second, we have discussed a secure an efficient scheme to reduce the overhead of managing the payment. Since the communication sessions may occur without involving a trusted party, the nodes have to submit the payment data to an off-line trusted party to update their payment account. Reducing the overhead of

M.M.E.A. Mahmoud and X. Shen, *Security for Multi-hop Wireless Networks*,
SpringerBriefs in Computer Science, DOI 10.1007/978-3-319-04603-7__5,
© The Author(s) 2014

submitting the payment data and processing them is essential for the effective use of micropayment for stimulating cooperation in MWNs. Because of using micropayment, most of the transactions may have low value and thus a transaction overhead in terms of payment data submission and processing should be much less the transaction value. In the proposed payment report based scheme, the nodes submit lightweight payment reports having their alleged charges and rewards for different routes, and temporarily store undeniable security *Evidences*. The fair reports can be cleared with almost no processing overhead and with acceptable delay. For the cheating reports, the *Evidences* are requested to identify and evict the cheating nodes. In order to reduce the *Evidences*' storage area, *Evidence* aggregation technique has been developed.

Third, we have discussed a mechanism to identify the malicious nodes that drop packets and break routes intentionally to launch *Denial of Service* attacks. Communication routes may be broken due to non-malicious reasons, such as mobility, or intentionally due to malicious actions. In order to differentiate between malicious and non-malicious route breaking, reputation system has been used to measure the nodes' route breaking rate. Two techniques, called *SRT* and *WRT*, have been used to calculate the nodes' reputation values. *SRT* offers equal negative ratings to the two nodes in a broken link, but *WRT* offers more negative rating to the low-reputation node that broke more routes in the past. With using the nodes' reputation values and several thresholds, the malicious nodes can be identified.

Finally, we have discussed routing protocols to establish reliable and stable routes. Multi-dimensional trust system has been used to measure the nodes' competence in relaying packets. The trust values can be computed with using contextual information extracted from the payment reports. The nodes' trust values are attached to their certificates to integrate the nodes' past behavior in the routing decision making. Considering trust in routing decisions is essential in the civilian applications of *MWNs* that is characterized by uncertainty in the nodes' behavior because they are autonomous and self-interested. Then, two routing protocols, called *SRR* and *BAR*, has been discussed to establish stable routes through the highly trusted nodes having sufficient energy to minimize the probability of breaking the route. *SRR* protocol establishes the shortest route that can satisfy the source node's requirements including energy, trust, and route length, and in *BAR* protocol, the destination node establishes the best available route.

The main benefits of integrating the payment and trust systems with the routing protocol can be summarized as follows. First, it fosters trust among the nodes by making knowledge about the nodes' past behavior available. Relaying packets by unknown nodes entails a certain element of risk, so a source node needs to trust the nodes that relay its packets. Second, this integration can deliver packets through reliable routes and allow the source nodes to prescribe their required level of trust. Third, it can stimulate the nodes not only to relay packets, but also to maintain route stability and report correct battery energy level. This is because any loss of trust will result in loss of future earnings. Fourth, the integration of the payment and trust systems with the routing protocol can punish the nodes that report incorrect energy capability. This is because the routes will be broken at these nodes and their trust values will degrade. Finally, a node may use a greedy strategy: never earn too much un-needed credits and stop relaying others' packets after earning sufficient credits. The integration of the payment and trust systems not only stimulates the nodes to cooperate in relaying packets to earn credits, but also stimulates the wealthy nodes to cooperate to maintain good trust values. This is because the nodes lose trust over time if they do not cooperate. By this way, in addition to payment, trust is another incentive for cooperation.

The robustness and the performance of these schemes and protocols have been evaluated with using simulations, measurements, analysis, and mathematical modeling. The evaluations have confirmed that the limit use of the public key cryptography can significantly reduce the overhead of the communication protocol. For a series of two packets, the protocol has lower cryptographic delay and energy than signature-based protocols, and for a series of 13 packets, the protocol requires around 10 % of the cryptographic delay and energy of the signature-based protocols. Moreover, the packet overhead is less than that of the *DSA* and *RSA* based protocols, e.g., for a series of 10 packets, the data packet overhead of our protocol is 70 % and 37 % of those of the *DSA* and the *RSA* based protocols, respectively. The evaluations have also demonstrated that the payment can be cleared with almost no processing overhead and submitting lightweight payment reports, and *Evidences* require low storage area. Moreover, *WRT* technique can precisely identify the malicious nodes with negligible false positive ratio because the honest and the malicious nodes receive less and more negative ratings, respectively. The reputation system is secure against small-scale irrational collusion attacks and robust against large-scale collusion attacks because the attackers lose credits and defame their reputations when launching these attacks. The simulation results demonstrate that *SRR* and *BAR* routing protocols can establish stable routes due to relaying the packets by the highly trusted nodes having sufficient energy, which can improve the packet delivery ratio.

5.2 Future Research Directions

Delay tolerant wireless networks (*DTNs*) [1–5] are an emerging class of multi-hop wireless networks characterized by long packet delivery delay and lack of fully connected route between the source and destination nodes. Consequently, packet transmission follows a store-carry-and-forward approach. The mobile nodes, acting as packet relays, buffer in-transit packets until the next node in the route appears, and so on, until the packets reach their destinations. Many useful civilian applications have been developed for *DTNs*. Pocket-switched *DTNs* take advantage of the increasing popularity of the mobile devices equipped with wireless network interfaces to enable a new class of social networking applications. *DTNs* can be readily deployed at low cost in developing and remote areas. Vehicular *DTNs* can be used for disseminating safety and location-dependent information. For mobile sensor *DTNs*, sensors are attached to the nodes, e.g., vehicles, to monitor the environment and the state of roads, e.g., potholes and black ice. However, the practical implementation of *DTNs* is questionable because the networks' unique characteristics have made them vulnerable to serious security threats.

The selfish nodes drop others' packets because packet relay consumes their resources without any benefits, and the irrational attackers such as compromised or malfunctioned nodes launch *Denial-of-Service* attacks by dropping the packets. In addition, due to the nature of wireless transmission and multi-hop packet relay, the attackers can analyze the network traffic to infer sensitive information such as the users' communication activities and locations. The privacy violation attacks can be launched in an undetectable way when the attackers just overhear the transmissions without disturbing the communication. These attacks are severe threat to the network proper operation and users' privacy. The presence of even a small number of attackers results in repeatedly dropped packets, which may cause network failure.

The protocols and schemes that have been discussed in this brief have been designed for the traditional *MWNs*, and cannot be used for *DTNs*. This is because they require establishing end-to-end route before the data transmission starts. Moreover, unlike the traditional *MWNs* that transmit one copy of each packet, multiple copies are transmitted in *DTNs* to enhance the packet-delivery probability. This will make the design of efficient incentive schemes a real challenge [6]. Involving many nodes in packet transmission will significantly increase the incentive scheme overhead in terms of the number of payment reports and *Evidences* and the number of cryptographic operations performed by the nodes.

In addition, in the traditional *MWNs*, each intermediate node is supposed to transmit a packet shortly after receiving it, and thus each node can monitor the transmissions of the successor node in the route to evaluate the node's trust value (or packet-relay probability). The trust values are used to identify the malicious nodes and relay the packets by the highly trusted nodes to maximize the packet-delivery probability. However, the unique "store-carry-and-forward" packet relay approach makes using this monitoring technique infeasible *DTNs*. It will also make the technique introduced in this brief to compute the trust values by processing the payment reports infeasible. Therefore, it is very important to study designing efficient and secure incentive scheme for *DTNs*. In this brief, we have introduced novel approaches to reduce the overhead of the incentive scheme for the traditional *MWNs*. This should be useful to devise new approaches for *DTNs* with considering their unique characteristics. In addition, designing a trust system to evaluate the nodes' competence in relaying packets should be interesting. In this brief, we have discussed a novel trust-evaluation approach for the traditional *MWNs* based on processing the incentive scheme's payment reports. However, applying such an approach to *DTNs* will face challenges, e.g., due to the multi-copy packet transmission. Designing routing protocol to relay the packets by the highly trusted nodes to maximize the packet-delivery probability is also very important. However, the difficulty is due to the probabilistic nature of the routing protocols in *DTNs*.

Finally, how to use privacy-preservation techniques such as mixers, onion routing and pseudonyms [7–12] to enhance the users' privacy in *DTNs* is very important.

The direct use of these techniques in *DTNs* may be infeasible because of the networks' characteristics. Integrating these techniques with trust-based routing protocol and incentive scheme should be carefully studied. In addition, providing high level of privacy preservation is important to make sure that any transmission in the network should not be linked to a node and cannot reveal the nodes' locations. On top of that, How to reduce the overhead is critical since most of the existing privacy-preserving schemes incur high overhead due to heavy packet broadcasting and public key cryptography

References

1. Z. Zhang. Routing in intermittently connected mobile ad hoc networks and delay tolerant networks: Overview and challenges. *IEEE Communications Surveys and Tutorials*, 8(1):24–37, 2006.
2. B. J. Choi and X. Shen. Adaptive asynchronous sleep scheduling protocols for delay tolerant networks. *IEEE Transactions on Mobile Computing*, 10(9):1283–1296, 2011.
3. R. Lu, X. Lin, H. Zhu, X. Shen, and B. R. Preiss. Pi: A practical incentive protocol for delay tolerant networks. *IEEE Transactions on Wireless Communications*, 9(4):1483–1493, 2010.
4. H. Zhu, X. Lin, R. Lu, Y. Fan, and X. Shen. Smart: A secure multi-layer credit based incentive scheme for delay-tolerant networks. *IEEE Transactions on Vehicular Technology*, 58(8):4628 – 4639, 2009.
5. P. Hui, J. Crowcroft, and E. Yoneki. Bubble rap: Social-based forwarding in delay tolerant networks. *Proc. of MobiHoc'08, Hongkong, China*, May 2008.
6. M. Mahmoud, M. Barua, and X. Shen. Sats: Secure data-forwarding scheme for delay-tolerant wireless networks, houston, texas, usa. *Proc. of IEEE Global Communication Conference (IEEE GLOBECOM'11)*, Dec. 2011.
7. M. Mahmoud and X. Shen. Secure and efficient source location privacy-preserving scheme for wireless sensor networks. *Proc. of IEEE International Conference on Communications (IEEE ICC'12), Ottawa Canada*, Jun. 2012.
8. M. Mahmoud and X. Shen. A novel traffic-analysis back tracing attack for locating source nodes in wireless sensor networks. *Proc. of IEEE International Conference on Communications (IEEE ICC'12), Ottawa Canada*, Jun. 2012.
9. M. Mahmoud and X. Shen. Cloud-based scheme for protecting source location privacy against hotspot-locating attack in wireless sensor networks. *IEEE Transactions on Parallel and Distributed Systems (IEEE TPDS)*, 23(10):1805–1818, Oct. 2012.
10. M. Mahmoud, S. Taha, J. Misic, and X. Shen. Lightweight privacy-preserving and secure communication protocol for hybrid ad hoc wireless networks. *IEEE Transactions on Parallel and Distributed Systems (IEEE TPDS)*, to appear.
11. M. Mahmoud and X. Shen. Lightweight privacy-preserving routing and incentive protocol for hybrid ad hoc wireless networks. *Proc. of IEEE International Workshop on Security in Computers, Networking and Communications (SCNC), INFOCOM'11, Shanghi, China*, Apr. 10–15 2011.
12. M. Mahmoud and X. Shen. Anonymous and authenticated routing in multi-hop cellular networks. *Proc. of IEEE International Conference on Communications (IEEE ICC'09), Dresden, Germany*, June 14–18 2009.